MOTHERSTONE

British Columbia's Volcanic Plateau

Chris Harris
Harold Rhenisch
Scientific Consultant: Dr. Mary Lou Bevier

MOTHERSTONE

British Columbia's Volcanic Plateau

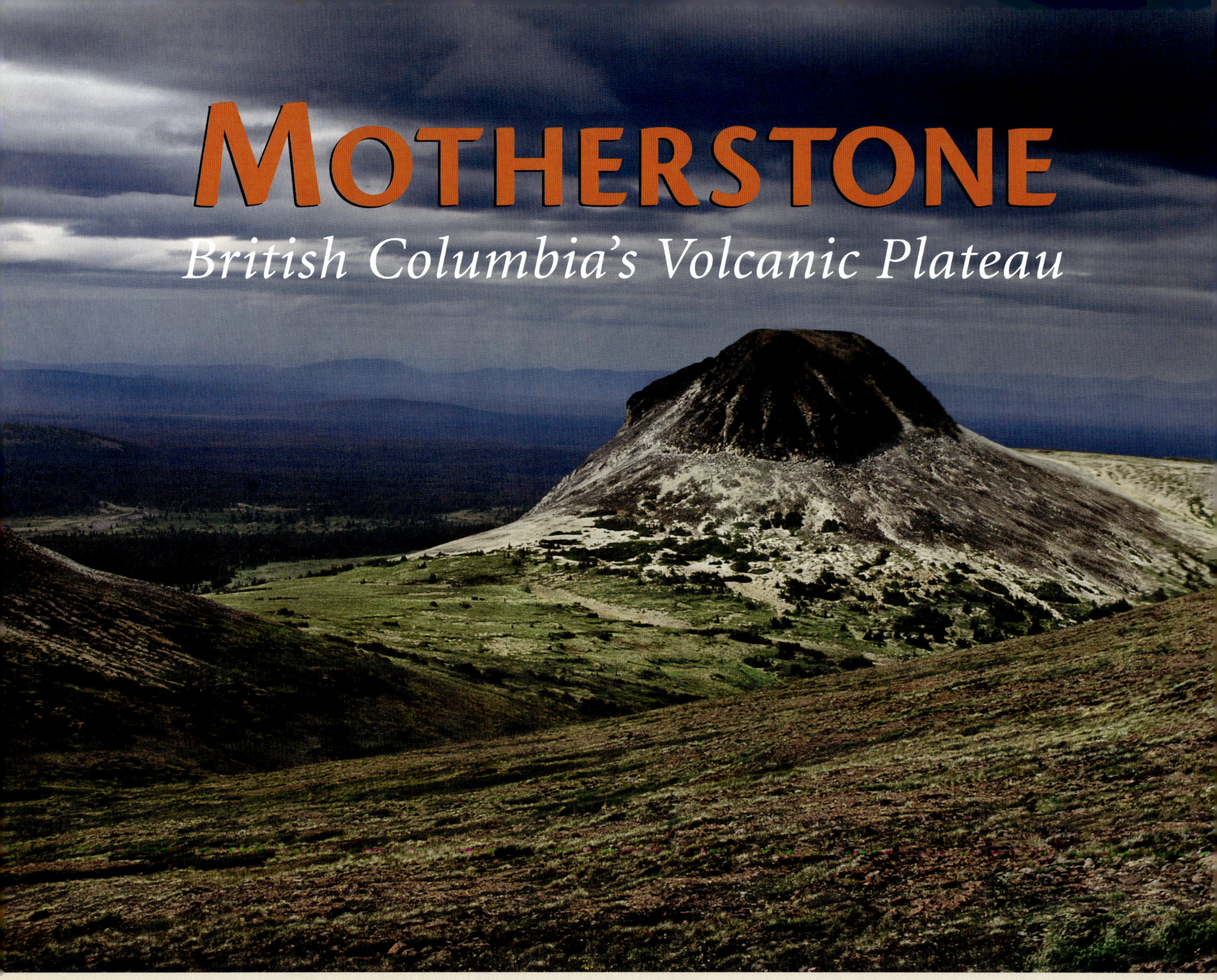

Photography: Chris Harris
Text: Harold Rhenisch

Scientific Consultant: Dr. Mary Lou Bevier

Country Light Publishing
105 Mile Ranch, BC

CANADIAN CATALOGUING IN PUBLICATION DATA

Edited by Harold Rhenisch
Scientific Consultant: Dr. Mary Lou Bevier
Designed by Bill Horne
Printed and bound in China

Library and Archives Canada Cataloguing in Publication
Harris, Chris, 1939-
Motherstone: British Columbia's volcanic plateau / photography by Chris Harris; text by Harold Rhenisch.
ISBN 978-0-9865818-1-6 (bound).--ISBN 978-0-9865818-0-9 (pbk.)
1. Volcanoes--British Columbia--Chilcotin Plateau. 2. Volcanoes--British Columbia--Chilcotin Plateau--Pictorial works. 3. Volcanoes--British Columbia--Cariboo Plateau. 4. Volcanoes--British Columbia--Cariboo Plateau--Pictorial works. I. Rhenisch, Harold, 1958- II. Title.
QE524.2.B75H34 2010 551.2109711'75 C2010-903269-1

All photographic images in this book have been made in-camera. Care has been taken to match this book's printed colours as closely as possible to the original film or digital capture.

PUBLISHED BY
Country Light Publishing

To order limited edition prints of any images in this book, or to order this book and others in the series *Discover British Columbia Books*™, contact:

Country Light Publishing
Box 333, 108 Mile Ranch, B.C.
Canada V0K 2Z0
T.250-791-6631, T.800-946-6622, F.250-791-6671
www.chrisharris.com
photography@chrisharris.com

Your comments are welcome.

Cover image: Pipe Organ Mountain. The heart of the Ilgachuz volcano.
Previous spread: The Itcha volcano is so remote and unexplored that most of its hard plugs and domes have never even been named.

Rita Giesbrecht chooses each footstep carefully on the first ascent from Carnlick Creek Camp onto the heights of the Ilgachuz volcano.

Through layers of crystalized basalt lava, Helmcken Falls plummets off the Interior Plateau on a line of gravity that flows directly to the centre of the Earth.

DEDICATION

I dedicate this book to Joyce and David Dorsey, and to Wanda and Roger Williams, who have dedicated their lives and passions to the three volcanic ranges of the West Chilcotin. They have carried with them the stories of the animals and the peoples who have lived in this area for millenia. They have fought for this land for years, and have kept it sacred through decades of governmental and corporate intrusions, and they opened the door for me. For fifteen years they have allowed me to share their lives, their stories, and their mountains. Their generosity has made this book possible.

Acknowledgements

This project is the work of many people. For her support, encouragement, intellect and vision, and for coming with me on many of my extended explorations, I thank my partner in life, Rita Giesbrecht. For his friendship, for walking with me and for contributing inspiration and energy, I thank Mike Duffy. For bringing his inspiration and deep land sense to the project, for his energy and vision as writer and researcher, and for partnership in shaping this project from its first to last moments, I thank Harold Rhenisch. For providing technical guidance as computers have become a dynamic part of the photographic process and for preventing me from becoming a technological dinosaur, I thank Shayne Middleton. These four people have walked every step of this journey with me.

Two people who have worked closely with us are Bill Horne and Dr. Mary Lou Bevier. Bill's masterly design work has brought the threads of this project into its living form. Mary Lou, the first scientist to document the volcanics of the West Chilcotin, has drawn in a community of geologists to ensure that our science remains accurate. It is an honour to work with such friends.

Another group of invaluable people are the guides and outfitters who took me into remote areas. Their support, warm-hearted friendship, and knowledge of the land were literally a lifeline. They include David and Joyce Dorsey of Rainbow Mountain Outfitters in Anahim Lake, and Wanda and Roger Williams of Itcha Ilgatchuz Mountain Outfitters in Six Mile Ranch. These guides, who have unconditionally supported all of my Chilcotin book projects, took me in by pack train to otherwise inaccessible areas. Their local knowledge was deepened by the generosity and skill of my friend Bill Mero, who took me onto the Itcha volcano by horse and showed me volcanic features I would never have found on my own. To all of these guide outfitters and their wonderful horses, thank you.

I thank as well my long-time friend Mike King of White Saddle Air Services, who provided photographic access to the most remote and spectacular regions of the Coast Mountains, and bush pilot Duke Sager, who flew me throughout the Rainbow, Itcha, and Ilgachuz volcanic ranges. The aerial perspective they provided was integral to the planning of this project, and provided some of this book's most stunning images. Access to other vital volcanic areas was provided by Dave and Rosemary Neads, who put us up at their home in the Precipice and sent us on our way to hike and photograph the historical Sugar Camp Trail to Bella Coola, and Brian Wear-Eylmann, founder of Lava Inc., who gave me full access to his property so that I could photograph his Nazko Cone operations, and guided me to a number of hidden and remote volcanic features that he had discovered over the years.

This project is part of a community of many such friends, who have supported it in a myriad of individual ways. They include Paul Adam, Graham Andrews, Karen Barger, Sage Birchwater, Ken Blackburn, Wolfgang Bogdanski, Mike and Darlene Calyniuk, Teresa Donk, Leslie Dorsey, Leslie Duffy, Tiffany Edwardsen, Tom, Jen and Sarah Ellison, Billy and Tiger Feistmantl, Kathy Fell, Arifin Graham, Grace Harris, Terra (Punky) and Eli Hatch, Dean Hull, Aileen Jorgenson, Dave Jorgenson and Cheryl Macarthy, Nicole Kipfer, Sandra Knueppel, Claire Kujundzic, Nomi, TalTal and Peleg Levi, Paul Lowrie, John and Evva McCarvill, Barb Mero, Jim Mortensen, Jeffrey Newman of JN Web Design, Jane O'Malley, Stanley Provost, Diane, Anassa and Leandra Rhenisch, Anna Roberts, Fritz Schaer of Wells Gray Air Services, Sherry Stewart, Stuart Sutherland, Jordon Tucker, Tanja and Brett Wagner, Anna Wärje and Richard Williams, Anita Willis of *British Columbia Magazine*, and Sandy Winston. To all of you, thank you. This book is ours, together.

As a photographer, I continually appreciate the inspiration I have received from my father, Chic Harris, my sister Jane O'Malley and her late husband Tony O'Malley, and photographers Freeman Patterson and Jim Brandenburg. Thank you, everyone!

Table of Contents

Anahim Peak towers over the West Chilcotin landscape at the edge of the eastern flank of the Rainbow volcano. Because it was a vital obsidian source, in its time this peak was as important industrially and militarily as the Middle Eastern oil fields are today.

The Motherstone

The rock in this book is a unique rock, lying on top of ancient volcanic islands in the Interior of British Columbia. On government maps, the area is known as the Cariboo-Chilcotin, but locally it is called the Motherstone. It lies between Anahim Lake in the west, the West Road and Horsefly Rivers to the north, the Cariboo Mountains to the east, and the Thompson Valley to the south. Here on this volcanic plateau of old basalt flows, fault lines, ghost rivers, shield volcanoes and cinder cones, on a planet almost completely transformed by human activity, the Earth's presence is still pure.

British Columbia, showing the Motherstone and related volcanic areas.

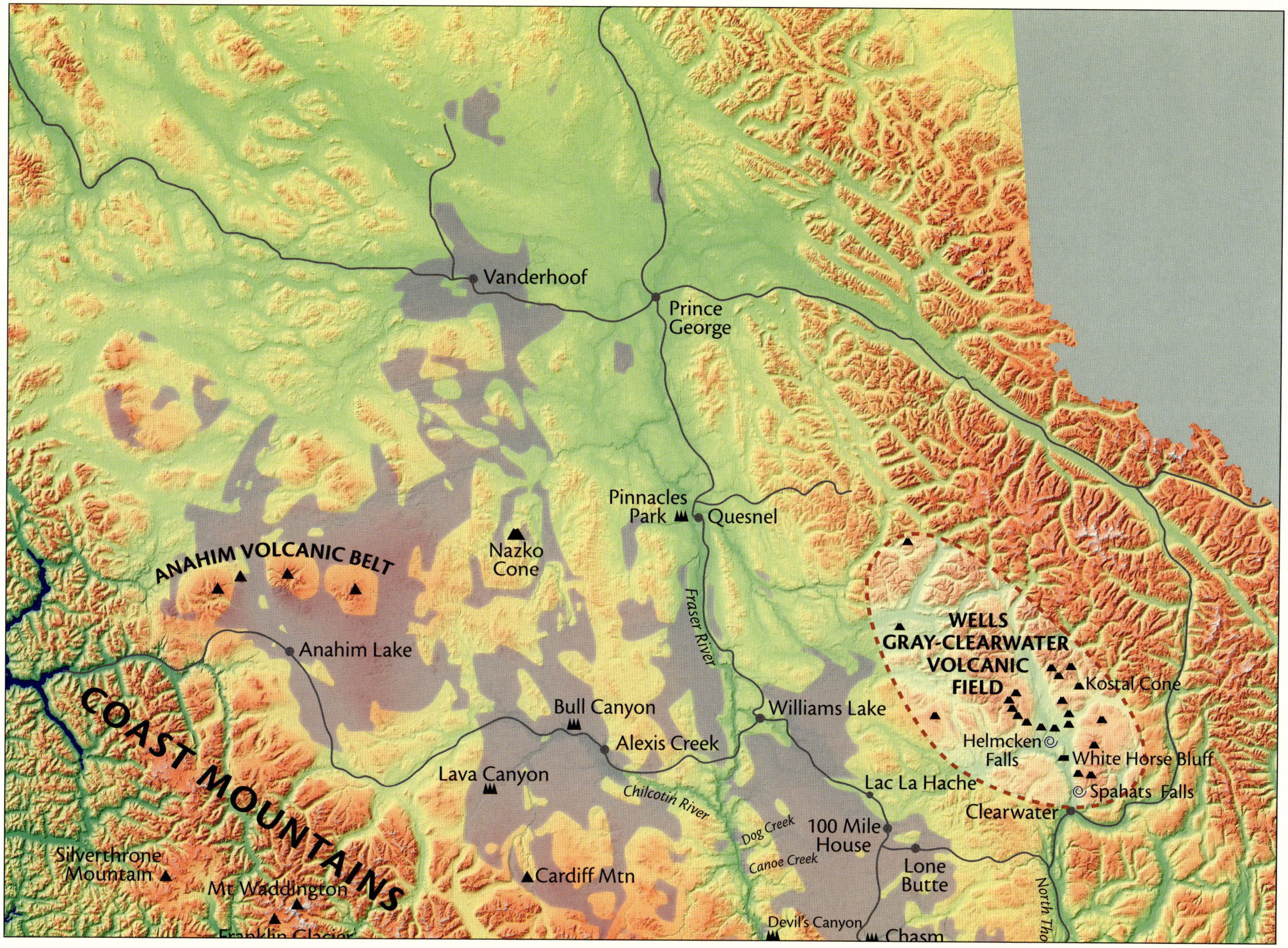

The Chilcotin Group basalts (*purple areas above*) lie like ancient seas upon the land.

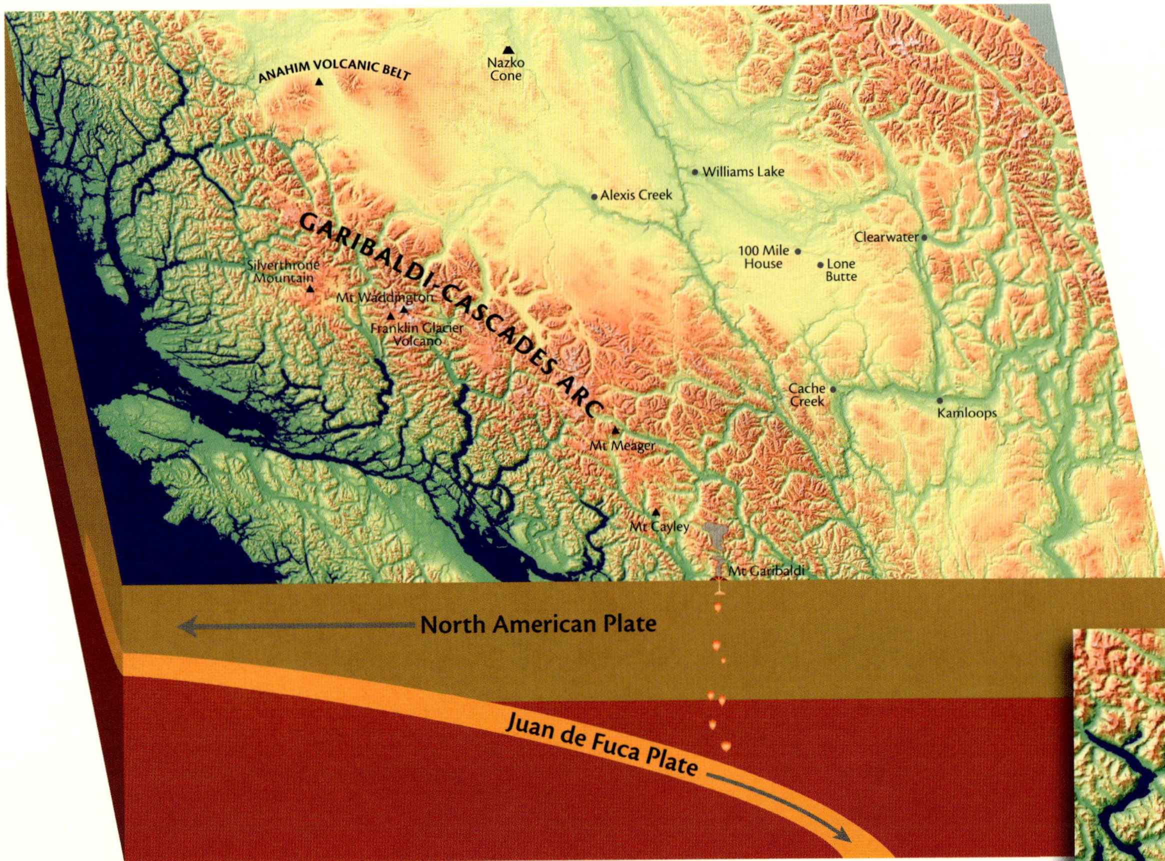

The Garibaldi-Cascades Arc

As the Juan de Fuca Plate subducts eastward from a line west of Vancouver Island and is pushed downward by the westward-moving North American Plate, it is subjected to extreme compression and releases water, which aids in the melting of the mantle above it. The magmas it helps form in this way rise to create the Garibaldi-Cascades Arc, an array of stunning stratovolcanoes that parallel the British Columbia coast, inland from the converging plate boundary. To the east, the basalt plateau of the Motherstone formed as a direct consequence of this activity.

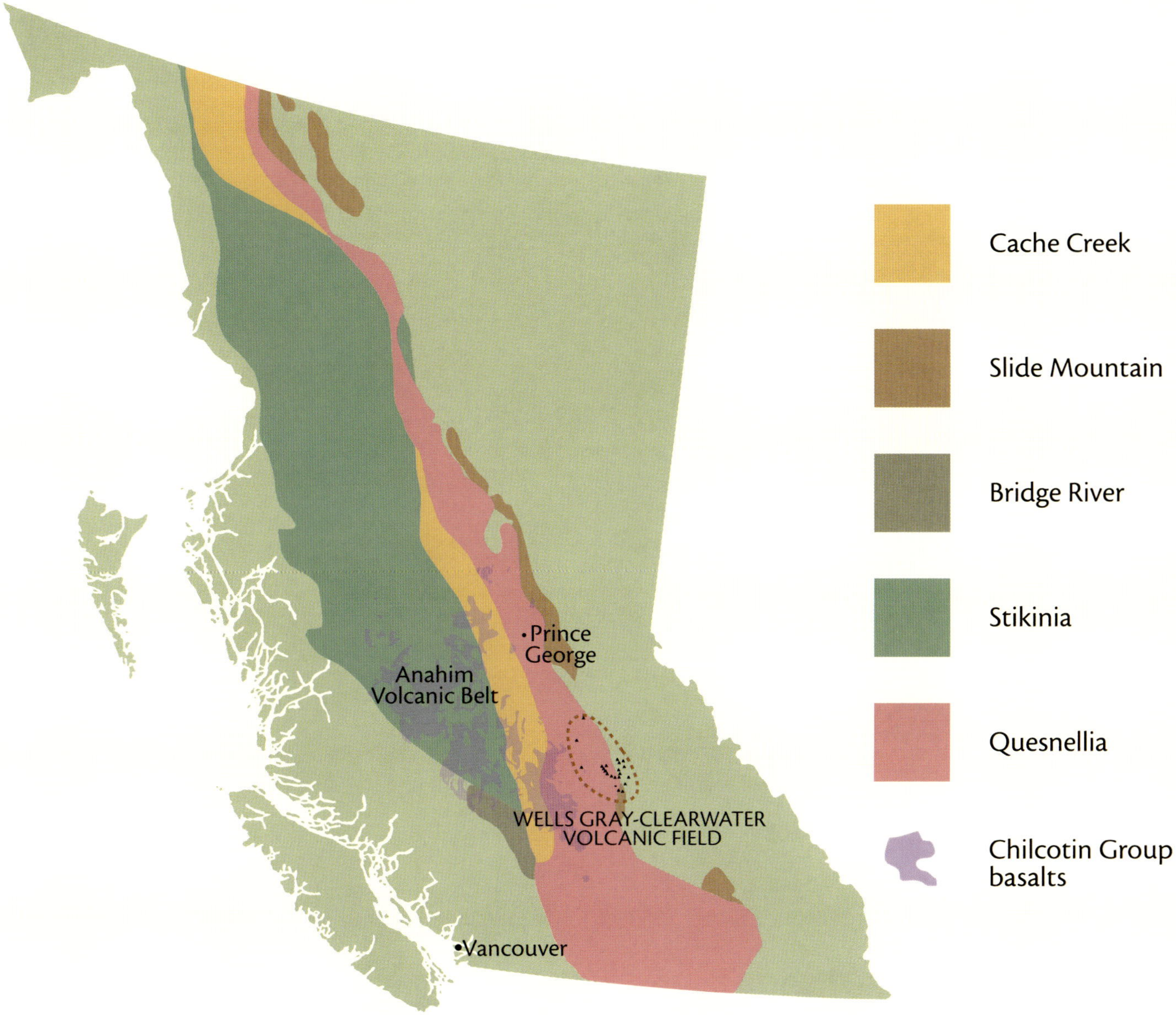

Ancient Terranes of the Motherstone

A terrane is a region of rock from a foreign source that has come a long distance before finding rest. Five terranes of extremely ancient islands from the far Pacific Ocean have welded themselves onto the coast of North America, and form the rock below the plateaus of the British Columbia Interior. They are the Cache Creek Terrane, the Slide Mountain Terrane, the Bridge River Terrane, Stikinia, and Quesnellia. Each has myriad local variations, and each is completely distinct from the others.

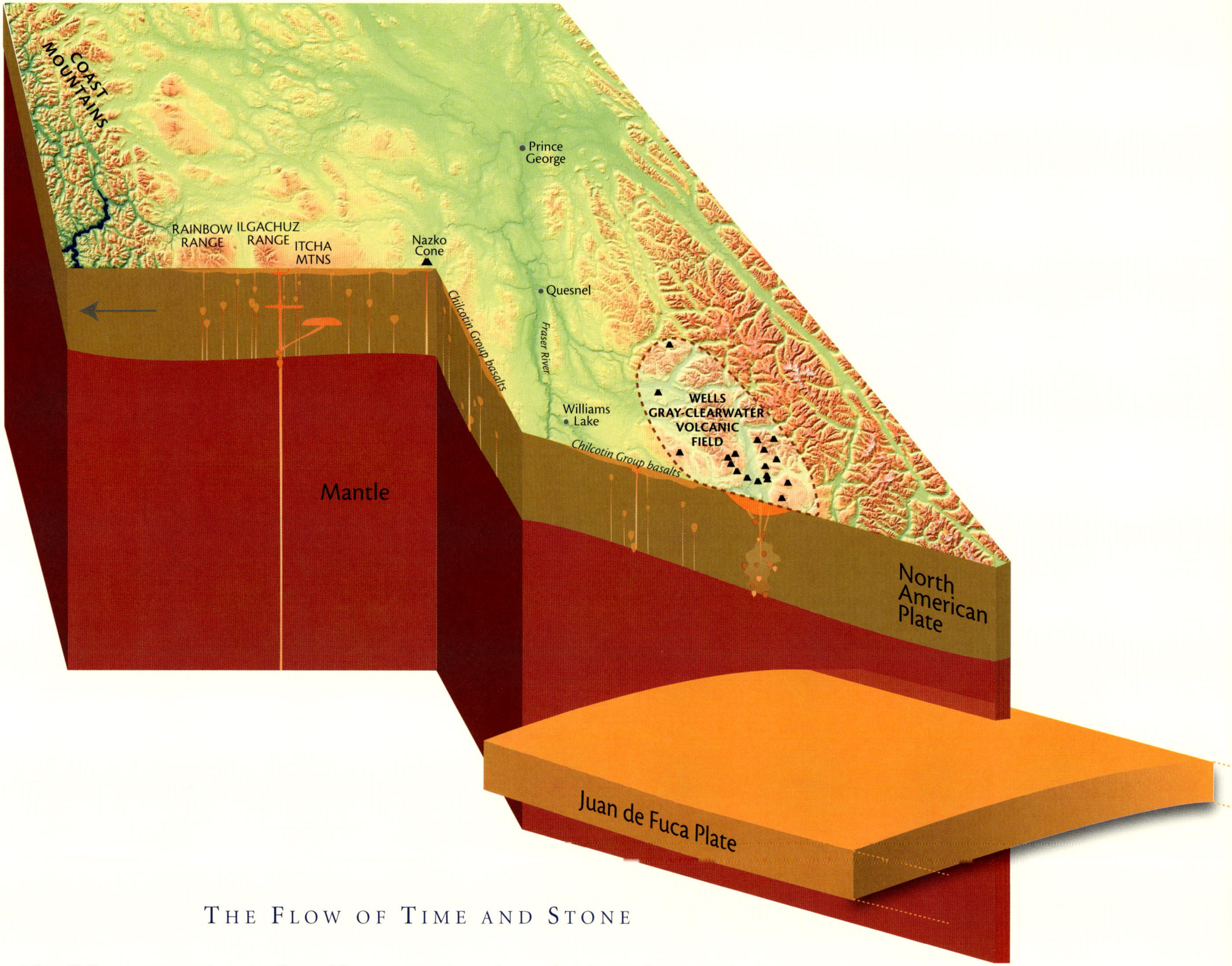

The Flow of Time and Stone

The Chilcotin Group basalts flowed from approximately 150 local vents between the Coast and Clearwater Mountains peaking around 10,000,000 years ago, at the same time that the Anahim Volcanic Belt formed as the North American Plate moved over a deep mantle hot spot. This image shows the hot spot under the Ilgachuz volcano 5,000,000 years ago. It is currently under Nazko Cone. Finally, the Wells Gray-Clearwater Volcanic Field rose among the ancient volcanic cores of the Clearwater Mountains, just north of the subducting Juan de Fuca Plate. Geologically, this area is still very active.

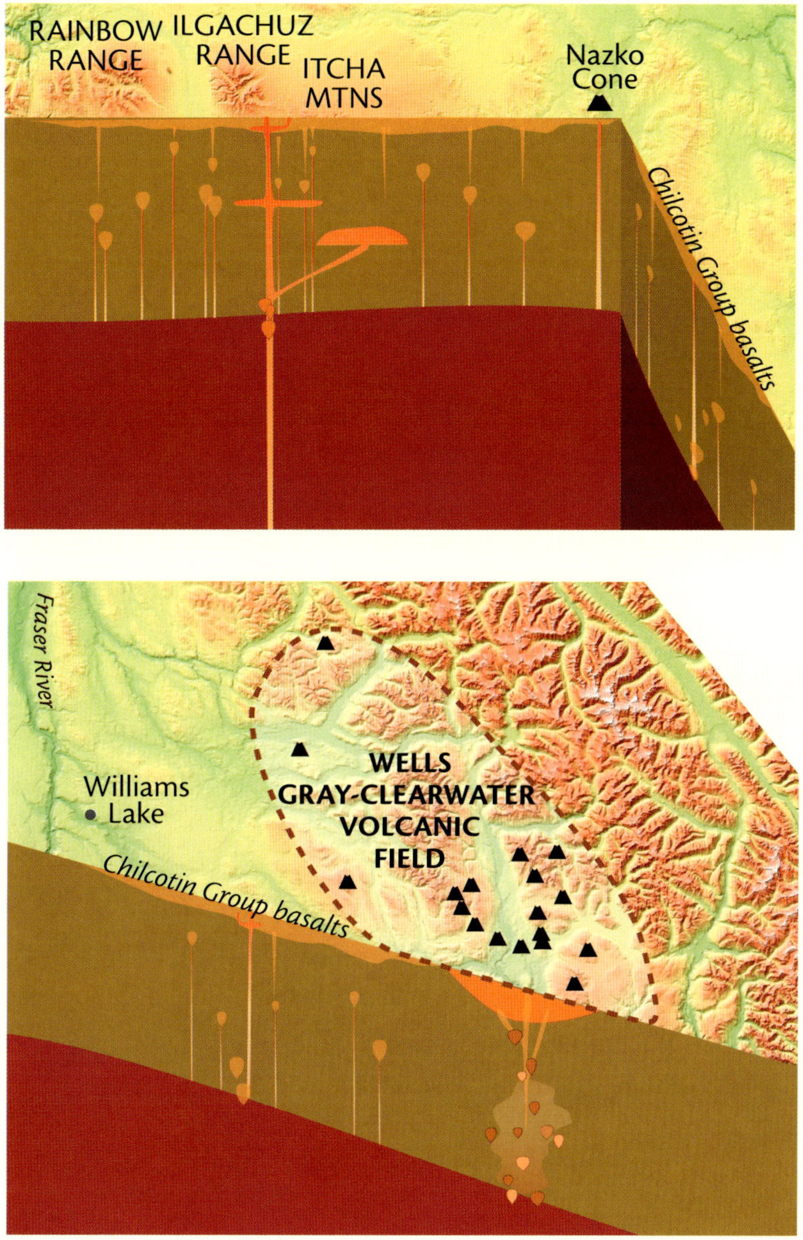

The lavas of the Anahim Volcanic Belt erupt from underground magma chambers into a series of shield volcanoes as the North American Plate slides west over them (*top*) and the lavas of the Chilcotin basalts that form the Motherstone plateau flow from the top of the mantle (*top and above*) to flood the land. Much later, the lavas of the Wells Gray-Clearwater Volcanic Field rise from just below the boundary of the mantle and the North American Plate (*above*).

The Volcanoes of the Motherstone

The Motherstone is formed from three distinct types of volcanoes, two of them related to the subduction under British Columbia of an ancient plate of the Earth called the Juan de Fuca Plate. The Chilcotin Group basalts flowed from many local vents as this land east of the Garibaldi-Cascades Arc volcanoes cracked when the subducting plate lifted the Coast Mountains high into the air. The volcanoes of the Wells Gray-Clearwater Volcanic Field have erupted in a tantalizing pattern in the area of a north-south trending crack in the Earth's crust called the North Thompson Fault and above the gap between the Juan de Fuca Plate and a fragmented remnant of it, called the Explorer Plate, to its north. Today, the Explorer Plate appears to be eddying deep beneath the Motherstone and moving so slowly it might not be moving at all. The third volcanic feature of the Motherstone is unrelated, although it too tracks the movement of the North American Plate. As the plate has moved west over an unmoving 2,800-kilometre-tall column of heat near the Motherstone's northern boundary, lavas that have collected in magma chambers deep underground have erupted one after another into a succession of low shield volcanoes, and, most recently, the Nazko cinder cone.

Cupped in a shoulder of the Rainbow volcano as it rises above the plateau to the east of the Coast Mountains, a flow of white stone lies as lightly as the last remnants of a late spring snow.

Geologist's Foreword

In 1976, aerial photographs of the Rainbow Range, a shield volcano that exhibited lava flows and glaciated valleys radiating from its core, enticed me to travel to the land Chris Harris calls the Motherstone. Why was a volcano there? How old were its lavas? First Nations peoples, who had lived at one with this land for generations and were aware of its character, had used its volcanic bounty — obsidian for spear points and red ochre for paints and pigments. But most travelers to the area remain unaware of its special geologic nature. I knew that unraveling the volcanic history of the area would increase public awareness of its uniqueness, so I strove to unlock the volcano's secrets. The immensity of the landscape and the elemental forces at play within it humbled me. By living within the volcano's inner valleys and traveling its ridges, enduring weather and bugs, and reveling in caribou sightings and mountain vistas, I slowly discovered how the volcano came to be. Thirty-four years later, we still know little about most of the volcanoes in central British Columbia. Through this book, knowledge of the region's volcanoes will reach a wide audience and increase their connection with this incredible area. It is my dream that *Motherstone* will inspire young people to study geology and lead them to new discoveries that will further unfold the story of this precious land.

— Dr. Mary Lou Bevier

Outfitter Roger Williams and his horses come face to face with the sky on the tabletop that crowns the Itcha volcano.

Even at mid-day alpine water reflects the deep indigo of the highest atmosphere, where it turns into the night of space itself. *Ilgachuz volcano*.

Caught by the Light

In 1996, I travelled by horse through the Ilgachuz Mountains. Everyone had their own reason for being there far off the beaten track. Mine was to capture images for my book *Chilcotin*. The plan was a good one: I would be back in the wilderness, where I have always felt most alive, and I would be making images in a mountain range I had only heard of before in the stories of my outfitter friends Wanda and Roger Williams. Little did I know how significant one of those images would be.

The custom on that trip was to take a nap after lunch. I've never understood why anyone would want to sleep in the midst of such beauty, so while everyone else settled down on a patch of grass, I took off with my camera. On one particular day I hiked up and over a steep ridge. Suddenly, I was staring down at what seemed like a volcanic crater with a small lake at the bottom. That image burned onto the emulsion of my mind. I promised that I would return one day to photograph there in greater depth.

Twelve years later, that image sparked the idea of creating a book on the volcanoes of the region. Once again I would get to visit the mountains and walk on cinders and touch cobbles and tumbled glacial stones that only wind, rain, and the feet of caribou, bears, and mountain goats had touched before me. So it began, as simply as that: my book about rock.

I knew few facts about the Motherstone at first, other than how the light struck it and how it changed into a new landscape with every drop of rain. I didn't know the names or histories of its minerals and basalts, so I started asking questions. Volcanologists, geologists, and botanists told me about pollen core samples, back arc volcanics, the Fraser Fault, the depth and extent of glaciation, and the ancient tectonic plates under my feet. They told me how the Earth's molten mantle had been cast out of the earth in fire, about mineral content,

Each of the tiny stones that make up the Rainbow volcano is made of a moment of time stretching back to the beginning of the world, and forward to its future.

Valley glaciers carved the Rainbow volcano into star patterns. They also revealed the succession of thin flows that formed it, like silt settling onto the bottom of the ocean.

The stone of the Ilgachuz volcano formed over a period of 5,000,000 years, was cut into intricate shapes by glaciers, and then was polished for 10,000 years by erosion and gravity. The photographer captures all these forces at once.

Unknowingly, Chris made two images of this moment fifteen years apart. In all that time, a few medium-sized rock shards and a tuft of grass had migrated about a metre downhill. In fifteen years time, the moment will still be developing.

oxidation, durability, and about the millions of years it took for the Motherstone to take on the shapes and colours it has today. As a photographer, I went further. I merged this data with what I knew about how a stone feels in the hand, how that looks to the eye, and how it feels when you stand there and the wind blows over you and smells of distant rain. I went to the mountains.

I travelled every way I could. In the end, plane, truck, car, and horse all led to one end: my feet. They carried me to the stone itself. Somewhere along the way I discovered that the Earth was alive around me — moving, disintegrating, and flowing with clouds, snow, streams, stars, a moon, the sky, rain, rocks, and grass. I felt energy that had once spilled liquid rock across the land flowing up around me still. As a hiker, I responded to it emotionally. As a photographer, I used my camera to explore those emotions, along with the stone and energy that gave rise to them. Over the weeks and months, as my feet carried me deep into geological time, I began to walk more and more in my own experience. I began to understand that the moment between the rock beneath my feet and my images of that rock was beauty.

In my excitement, I began to chase the light across the volcanoes. I became attuned to moments of balance, and kept my sight sharp by always looking for points of surprise. The mountains did not disappoint. As clouds disappeared into open air around a peak or a mountain suddenly flared in front of me as I rounded a spur, it seemed like I was chasing an elusive free spirit that always revealed itself when I least expected it. I learned not to expect it, to wait for it to come to me, to be forever ready. Sometimes while I was enduring the chill of crossing a mountain stream, or when I was on a tough slog through marshy meadows or navigating through a whiteout by compass and memory, I would feel suddenly alive, and my heart would soar. When I was lucky, I was able to turn such moments to art.

My photography is a form of ethics, of respect for the enormous fund of energy given freely by our planet. This planet is life. It created these sculptures. My job has been to see them, and then to go beyond seeing and to set you on the same journey into light. To do this, I have used the simple palettes of form and light to paint the unseeable that the Earth expresses as art and we all express as joy, pain, wonder, cold, and memory. I have tracked the strong, solid shapes of the mountains and have kept myself alive to subtle tapestries of texture, colour and tone. I have made images without sense of scale, to portray my moments of discovery. I have followed my elements into compositions within compositions within compositions, and have found smaller sculptures, and smaller paintings, and infinite combinations and recombinations, within every mountain sculpture and every painting glowing from stone. Welcome to this journey.

— Chris Harris

A view of the land into the distance is not necessary when a view into the land is possible in its place. The glaciers have long since stripped away the old cinder cone of Mount Downton in the Itcha volcano and carried it south-west towards the Pacific Ocean, but its hard core and the golden fields of cooked stone remain, turned into gentle folds by seas of ice. Here it is possible to walk right up to the face of the sun, and to touch it.

This rich palette of shape and colour in the Ilgachuz volcano was laid down when volcanic vents lying side by side erupted at different times with different mixtures of lava cooked underground for varied lengths of time. When Chris found these mountain colours, the sun was still adding its heat to the mix.

Art is the moment when a photographer sees the mountain within the mountain. The sky reaches down and touches the ground. As we watch, we hold our breath, yet our breath still blows.

The camera follows Mike Duffy on his descent into the centre of the Rainbow volcano.

Finding the Motherstone

For two centuries now, we have all been on an amazing spirit journey to the beginning of time. Scientists have shown us our universe when it was smaller than a flake of granite, and have written mathematical equations for the bands of energy inside the smallest grain of sand. They have demonstrated how this dance had formed from the dust of a long-exploded mother star, in a galaxy formed from many other long-exploded stars and galaxies, rising and falling in the surf of time.

For many of us on this surging sea of stone, home is a lakeshore on a high plateau lifted between 1,000 and 1,700 metres into the sky and falling back to the sea like a wave on a long shore facing the open Pacific. In the spring, the lakes scattered across this wave hold all the light in the world and the whole planet seems like a sphere of light over which the lakes themselves are only cracks that allow it to escape, briefly, into space.

This is the Motherstone. It is a big country that contains the last two pristine temperate grasslands on our planet, the last big herd of mountain caribou, and, at its western edge, some of the last land on Earth that has never been walked by human feet or seen by human eyes. Over that edge is the white surf of what may well become the last glaciers. To walk here is to truly see the beginning of the world, the moment when it was brought to light and words were found for it, like a match struck in darkness.

Chris Harris, Rita Giesbrecht, and Mike Duffy found this story simply, in the oldest way of telling any tale: by walking. I came in at the end of their first walkabout, four years ago. In the early spring of 2006 I went to Farwell Canyon with Chris and the ecologist Ordell Steen and bent down and touched the soil between the bunch-grass. When I stood again, on a sun-warmed glacial erratic, I saw it as an

Chris set an image of red-hot coals and these rocks side by side in his darkroom, and the rocks flickered, and gave off heat. In that moment, the text of *Motherstone* began.

Rocks are burning around us all the time. Most often their coals burst into flame as flowers and grasses, such as these forget-me-nots rooted in the moisture of the previous winter's vanished snows.

island of stone in an ocean landscape. It was there that day, above the Chilcotin River, that I began this book. I talked to Chris and Ordell about islands of grass and stone and snow-shadow that could all be read in the grass. After a year of hard work, often late into the night, the result was *Spirit in the Grass*. Creating that book with Chris changed my life. I had been a man of the wind. I became a man of the grass.

There was no place in *Spirit in the Grass* for the ocean shore I saw on the plateau that day, yet it stuck with me, as did the feeling of standing in the grass and salsify at the side of the Riske Creek road and looking up to the grey rock of the ancient seabed atop the grassy hills and feeling as if I was standing in the trough of a swell.

As fate had it, within a year social pressures from the evolving story of the British Columbia economy drove me away — or so I thought. I packed up my home in the savannah of the old East Meadow of the Jones Lake Ranch near 150 Mile House and moved to Campbell River, on Vancouver Island, where the only grassland is a tiny strip of saskatoons and wild roses on the spit separating the river from the tide surges of Discovery Passage. Other than that, everything around me was cedar trees and tides, and rain.

Or so I thought. When Chris, Rita and Mike went out walking again, I was just a short distance to the west, picking my way through the boulders along the low tideline in Campbell River. I realized that day that we were both moving through the same landscape — a vast ocean shore stretching hundreds of kilometres inland from the sea, with a rainforest in the west, a country of dunes and draining rains and salmon to the east, and east of them both another rainforest, and beyond that, finally, the mountains and North America. I had found the shore where the Pacific Ocean and North America meet. It contains all the forests and mountains of the British Columbia coast, most of the drainage of

On the Ilgachuz volcano, eroded plugs and channels of stone attest to the extreme complexity of shield volcano eruptions that last for millions of years. In such complex structures…

…vents only a few metres apart can rise from different areas of the crust and mantle beneath them and from different reservoirs of molten rock as the continent moves westward over the hot spot.

Confronted with natural processes, early scientists deduced principles that they called the laws of nature. One of them, the Law of Entropy, states that everything loses energy and falls to a state of balance with its surroundings. Plants colonizing points of balance in these eroding slopes are also records of energy. They do, however, radiate light along the way, and we do turn that light into spirit. That is not a falling; it is a lifting up.

the Fraser and Columbia Rivers, including the Columbia Basalts, the Similkameen, Okanagan, Kettle, Thompson and Kootenay valleys, and, of course, the Motherstone of the Cariboo and the Chilcotin.

Within the next months, Chris and Rita and Mike made a series of trips through the shield volcanoes of the West Chilcotin, learning the story of rock and the story of the climax grasslands of the Itcha and Ilgachuz volcanoes. I was learning the story of water. Like a salmon, I was following my first river, the Columbia, home from the sea. Four months later, Chris, Rita, Mike and I sat down again in Chris's darkroom, and went through images of mountains and told stories until late in the night and again all of the next day. Anyone who came through the house sat with us for a while and shared ideas. I read out my first sketches for the text of *Motherstone.* Chris and Mike filled the living room floor with rocks. Chris and Rita's friend Nomi cooked sumptuous meals for us, so that we could continue talking, and when I left I had a bag of rocks in my car and the beginnings of a name for the land.

Since few people, including few aboriginal people, have ever been on the Motherstone's highest washes and bare passes, the Ilgachuz and Itcha and Rainbow volcanoes, there were simply no words for this land. There was only what we knew in our bones, only the feel of this unique country and the images Chris had used to capture it. At first, we fumbled for words. At first, down in Chris's darkroom we all thought, no, it can't be, this is not the Earth; we are on Mars, or Venus, or on some distant planet among the stars; and then we realized — together, I think — no, we live on a planet just like that, and it looks like this; this is what it's like to live among the stars.

The land does something to you out here. The stars of the Motherstone are like no stars anywhere else. The rock underfoot feels like it is floating — as if it were air or a boat at sea. The newest

All humans share the memory of refuge that our ancient hominid ancestors remember through us: a view from a cliff to trees along water. This image of the Myrtle River Valley brings us that history in an instant as brief as the click of a photographic shutter. Any trail we could follow down into the valley would also be a trail into our minds, our DNA, and the 200,000 generations we represent in the present of an unbroken story.

Everywhere, water follows ancient paths of energy. As the Earth cracks with stresses many kilometres below it, Spahats Falls spills through the flows of ancient stone that followed its path long before it.

Every year, the Rayfield River carries a fresh stream of the previous winter's snow into the old volcanic vent of Crater Lake, south of 100 Mile House, and pulls much of the air along with it. Even the poise of snow here radiates spirit.To stand above these Motherstone lakes is like looking directly into the Earth's molten core.

As it enters the Thompson Valley, the Deadman River slips through the channel cut by the meltwater torrent that left the towering alluvial fan seen at the upper right of the photograph above. Salmon fossils from the refuge of Glacial Lake Thompson, which once filled the Thompson Valley in the background, have been found within sediments to the west of this formation, just east of Cache Creek.

maps of this rock are created with the detailed measurements of seismographs and satellite imagery. We have included them here, in a place of honour. The oldest maps, however, are stories of the dreamtime, when the Earth was alive and all creatures upon it were people. *Motherstone* is written with deep respect for the knowledge that such stories contain. These stories remember the first days of our people — not the people of British Columbia or the Cariboo today, or even the Secwepemc or the Tsilhqot'in or the Carrier, but all the people who come from this land and cannot leave it. We remembered the time when there were no humans, no animals, no plants, and no rock — when we all moved through our days together.

— Harold Rhenisch

Many of the goats, caribou, wolves and marmots of this country have never seen people before. This mountain goat sure has that look about him. Here, we catch him looking into our world, while we look into his.

The ancient Greeks told the story that writing was invented by men reading the patterns of bird tracks on sand. Obviously, here in the Ilgachuz volcano it has been invented by goats.

To ride up through eskers in the graveyard of the glaciers in the Itcha volcano is to feel the streams that ran beneath the ice sheets flow around you in cold eddies and sudden, chill currents. With every footstep of your horse, the land of fire springs to life beneath you again, out of the water and the cold.

Rivers of Stone

Geologists teach that the Earth is hot. Its core conducts energy outward at more than 5,000° Celsius — the temperature of the surface of the sun. It's as if the sun were burning under your feet.

Now step outside and close your eyes for a moment — and look up. Perhaps it's a cloudless day. The sun is high. Feel its heat on your face, brushing your cheeks like a wind. Let it soak in for a while.

The energy in that warmth has come 146,000,000 kilometres through space that's holding at about 273 degrees below zero. Feel it. It ruffles your hair. Its heat soaks into you as you stand within it. You are living in its flame.

Now, imagine it's winter. You've put on a coat and are bending down to pick up a rock from your garden. The rock is cold. You have to scrape the snow off to get to it. You have to pry it out of the ice. Your fingers are raw and chill.

Feel that. This is your planet. Your body is made out of such rock. It knows that 5,000 degrees of heat that has come a mere 5,150 kilometres through warm rock from the Earth's core has vanished, and has left you holding the cold of the universe. It's as if you were in Williams Lake, British Columbia, and the burning sun, 1,392,000 kilometres in diameter, was plunked down only 5,200 kilometres away at the mouth of the St. Lawrence River, yet all that heat never made it to you at all.

Now, bend down and pick up that winter rock covered with snow and ice, and feel it — really feel it.

It is not rock your hand is holding, at all. Your body may feel weight and texture and temperature there, but it's still not rock — not really. Rather, it's the memory of a time when deep, underground pressures cracked the Earth and hot, liquid stone flowed over the land. The pressure was eventually relieved and the cracks eventually sealed, but the stone remained. Its flows are caught forever in the moment of their creation. They are a past that you can walk through and pick up and put into your pocket.

If you pick up a stone in 100 Mile House, you might have as much as 10,000,000 years in your hand. If you

Orange Creek carries the last of winter down through summer's grasses. *Ilgachuz volcano.*

The Rainbow volcano is a liquid mountain born in fire, that starts flowing again at the slightest drop of rain.

In the shield volcanoes, the forces of convection that shift the plates of the Earth are mirrored in the flow of gravity that moves crumbling rock like water and makes Niagara Falls look like a trickle in the spring.

drive down into the Fraser Rift near Big Bar, and pick up a green stone in Devil's Canyon, you will be holding 350,000,000 years in your hand. You could sit down for a moment and watch the waters flow, carrying all that time down to the sea.

All of us, together, are energy, on a planet that is a field of energy, compressed by gravity, spun into a sphere, and orbiting a yellow star. On our northern winter horizon, the magnetic fields of this energy flare up as the Aurora Borealis, rippling across the sky in sheets of red and green fire as if there were no line between water and air.

Deep within the Earth, this same energy flows in convection waves, like the bands of heat around a candle flame. When it strikes the surface of the sphere, where the Earth has chilled into hard plates of rock, it is deflected sideways and spun back to the Earth's core to be heated again. That motion of heat rising from the Earth's deepest pool and returning to it is also who we are. That energy is your hand that picked up that rock. That rock is also that energy. That energy is also your mind, which knew how to imagine it.

The scientists have imagined this for us: we live on plates of rock pushed around the surface of the planet like ice sheets on a high country lake. When the plates touch, they are forced under and over each other. The ones that are driven down — sixty, eighty, one hundred kilometres or even more — are dehydrated by the pressure. The water squeezed from them melts the Earth's mantle above them into magma that spills up in arcs of volcanoes inland of the line of collision. The arc closest to us runs through the Coast Mountains and down into the Cascades, on the northeastern coast of the Pacific Ocean. It includes Mount Baker, towering over the Fraser Valley; Mount Garibaldi, high above Vancouver; Mount Rainier, soaring over Seattle; and almost eighty others.

The coast they form against took shape 750,000,000 years ago, when the Earth broke apart in heat that poured from her iron heart. Earthquakes shattered what would eventually become the mountain homeland of the Nez Perce, Coeur d'Aleine, and Kootenai peoples. After long, slow journeys across the planet's surface, the Earth's plates were drawn deep under the continent again. They formed coastal arcs and mountain ranges and lifted plateaus behind them, and they did it time and time again. Over the past 400,000,000 years in British Columbia, four different ranges of volcanoes have collided with the coast of ancient North America in just this way.

Then the stone came alive. Life was born from the sea and found the land. Flowers bloomed, trees reached to the new light, and animals called into the silence and were heard. With them, the Earth grew green and fruitful, the air flooded with oxygen, the sky turned blue, yet still the Earth's iron heart kept burning, and it broke her into pieces like spores cast from a fern, and once more she drifted onto the sea, until each part of her was a separate continent, with its own animals, trees and flowers.

With life came healing. A huge sheet of rock called the Farallon Plate broke up under the Pacific Ocean 205,000,000 years ago. One after the other its fragments slid under North America. Since volcanoes at the edges of these fragments were too hard and dense and far too big to get sucked under as they splashed onto the shores of North America, the Earth rose up and welded them to herself. Then the next wave of islands came, and the next, and the next. They came like canoes to a potlatch, and together they made the lands men would call Washington, Idaho and British Columbia, which are moving west with the continent now, about three centimetres per year.

In British Columbia, these exotic volcanic terranes include the Cache Creek Terrane, the Slide Mountain Terrane, the Bridge River Terrane, Stikinia, and Quesnellia. You can walk on those chains of ancient islands. Chances are, you have already, perhaps for your entire life. When you're in Churn Creek, in the grasslands along

As mountain cinders flow with snowmelt and rain down to the sea, life meets them on its slow climb up to the sky.

In the Rainbow volcano, cinders from various eruptions can easily be distinguished by colour. Each colour of rock has a completely different texture. In these waves of stone, we can find where we are by our hands.

Billions of years in the future, the convected heat within the Earth's mantle and outer core will still. The plates of stone covering the Earth will lock together, and no more volcanoes will erupt, no new mountains will be born, and old ones will stop their slow rise into the sky. That is not our story. Our story is in the time of creation, now, when the Earth is alive, and new mountains are born. Chris took this birth picture in the Rainbow volcano, when the volcano was still so new that he could still trace its first flows.

the Fraser River, you're standing on the Cache Creek Terrane. When you're in Nimpo Lake, in the West Chilcotin, you're standing on one of the islands of Stikinia. When you're in Lac La Hache, in the heart of the Cariboo, you're standing on the eroded oceanic slopes of Quesnellia. This whole area is the salmon's inland sea, where the Earth's only inland temperate rainforest rises on the edge of a vast beach landscape stretching for hundreds of miles. The older part of North America lies far to the east, in Alberta.

The collision is still going on. The Juan de Fuca Plate, a fragment of the Farallon Plate, is subducting to the east, southward of a line between Nootka Sound on Vancouver Island, Bridge Lake in the South Cariboo, and the Rocky Mountains. A smaller fragment north of it is called the Explorer Plate. It is moving northward, but so slowly that it is almost stalled — perhaps it has begun to rotate on its axis, back towards the west. Where the gap between these two plates lies below the North Thompson Fault, the Earth has erupted into a ring of volcanoes called the Clearwater Volcanic Field. These are the sub-glacial volcanoes of Wells Gray Park — Kona, Dragon, Pillow Creek, Gage Hill, Ray, Spanish Mump, Hyalo Ridge, Mosquito Mound, McLeod Hill, Jack's Jump, Fifty-Two, White Horse, and Pyramid. A mere 400 years ago, they were joined by their youngest sister, Kostal Cone.

As the Juan de Fuca Plate continues its journey eastward, the Explorer Plate continues to fill the gap left under the Cariboo and the Chilcotin. In the process, it continues to lift the land into the air. So far, centimetre by slow centimetre it has managed to lift the entire centre of the province 600 metres. The Fraser and Thompson rivers lie at the bottom of this rise. To descend their banks down to the fine silts of their shorelines is to travel deep into time — perhaps as much as 150,000,000 years. On foot, it might take an hour.

Imagine you are standing on the bed of the Fraser, with floodwaters pouring past your feet and the horizon far above you, almost obscured by sun and sky. And on the riverbed there is a thin layer of red rock. That rock, scientifically named the Chilcotin Group basalt, is known locally as the Motherstone. When the underground lava chambers of ancient volcanoes were heaved into the western sky to form the Coast Mountains 31,000,000 years ago, the land was pulled like taffy up onto their slopes and broke under the strain. During the next 30,000,000 years lava flowed through the cracks — basalt as thin as honey. It spilled out in multiple flows, each cooling and layering upon the last. Lava rivers flowed quickly down ancient valleys until the valleys were full, then spilled over the river banks and the hilltops and created the plateau. That is the bed of the Motherstone — a pool of stone between the Yalakom and North Thompson faults, between the Coast Mountains and the Clearwater Mountains, interrupted only by the deep valley of the Fraser Fault. It lies above it, light and airy like sea foam, on the lavas of ancient islands. In some areas, her flows are hundreds of metres thick.

It is easy to see the path of the valley glaciers as they flowed down off the peak of the Rainbow volcano. When you are in this high country, the road of the glaciers is also yours. There is no other way to go.

The basalt of the Ilgachuz volcano is light with trapped air and easily shifted by water. These unique rivers of stone at the top of Orange Creek have the perfect mass to float down on snowmelt. Elsewhere on the volcano, lighter and heavier basalts containing different volumes of air react to the weather with their own unique signatures.

In the volcanoes, everything from soot so fine it feels like powdered sugar to boulders the size of mountains are textured and in various degrees of disintegration. These volcanic rocks on Mount Downton have broken up over the last 12,000 years and are beginning to flow again, set in motion by snowflakes.

In others, they scarcely conceal the region's ancient valleys, hilltops and streambeds.

The Motherstone's crowning feature rises above this lava sea. It formed 8,700,000 years ago, as a volcano spilling above the northern edge of the Explorer Plate — a low shield volcano, like Hawaii or Maui. She erupted in thin flows for 2,000,000 years from hundreds of domes, cones, and vents. She is the Rainbow volcano and she is 30 kilometres in diameter. 700,000 years after her eruptions ceased, when the continent had slid even further west, a second shield volcano formed. It too flowed for 2,000,000 years and also erupted from hundreds of different openings springing from a central magma chamber. We call her the Ilgachuz volcano. She is 25 kilometres in diameter.

A million years after the Ilgachuz volcano stopped flowing, a third shield volcano formed. This one flowed intermittently for almost 3,000,000 years — most recently 9,000 years ago, at the end of the last glaciation. She is called the Itcha volcano, and she is 10 kilometres in diameter. Her rocks are red and purple and gold. Together, these volcanoes are known as the Anahim Volcanic Belt.

It is commonly believed that the volcanoes of the Anahim Volcanic Belt erupted one after another from a phenomenon called a hot spot, in which a pipeline of heat leads directly to the Earth's surface from the core-mantle boundary through one unchanging point, melting the deep rock and pushing lava out in gentle flows. The islands of Hawaii stand above such a hot spot — or rather, they did while being formed; eventually, one by one, plate motion has pulled them away, leaving space for a new island to form behind behind each one in turn. Hot spots are ancient, perhaps as ancient as the Earth itself, yet the volcanoes of the Anahim Volcanic Belt are incredibly new. When their story is finally understood, it may startle us.

We do know a lot, however. We know that the plates are still moving. Between the Itcha volcano and Quesnel, for example, there is a fourth volcano. She is Nazko Cone, and her last major eruption was 7,200 years ago. Most recently, on October 10, 2007 a year-long series of

In the Coast Mountains, a plume of granite rose through older, darker country rock 80,000,000 years ago, as the edge of the North American continent surged over the northeastern edge of the subducting Kula Plate. Glaciers flowing west from the Motherstone sliced down through the uplifted rock to form the mountain we see today.

An Ecological Reserve

Once-level volcanic table lands, eroded in the Nemiah Valley into the bluffs called Cardiff Mountain, are being slowly tilted to the northeast as the land is drawn up into the slopes of the Coast Mountains. This mountain's rocky crevices host a unique fern that has adapted from its original cool, moist environment to an open, dry, and windy one. They are the only basalt columns in the Motherstone painted richly with orange lichens. The bottom of the lake to the right is covered with aquatic moss. Cardiff Mountain has been protected as an ecological reserve since 1974 and belongs to the creatures of this earth. It is off-limits to Homo sapiens.

The Deep Earth Among the Trees

Nazko Cone erupted in an explosion of black scoria and ash 7,000 years ago. A second explosion blew out the south side of the crater and freed the way for a 1,500-metre-long tongue of lava to spill into the forest. Chunks of air-cooled basalt called lava bombs still lie where this cataclysmic explosion threw them. Because of recent earthquake activity deep underground, Nazko Cone is currently listed as Canada's most potentially active volcano.

earthquake swarms signalled a movement of lava twenty miles to her west in the upper reaches of the Bazaeko River. As you are reading this, an underwater mud volcano called Mount Maquinna is building along the Nootka Fault on the southern edge of the Explorer Plate, seventeen kilometres west of Vancouver Island, as seawater pours into the expanding crack between the Explorer and Juan de Fuca Plates and is squeezed back out. Only thirty-seven years ago there was a series of earthquake swarms under Mica Dam, to the east of Clearwater.

The Motherstone will stop no time soon. Somewhere, near Nazko, perhaps, or Clearwater, or even near the Mica Dam, someday, perhaps in a decade, perhaps in 20,000,000 years, mother stone will again flow from the Earth, and again will burn the forests and fill the valleys and the air with ash and cinder. There is nothing stable about this land. It is a wave, breaking on the shore of the stars.

An Explosion of Colour

The palette of Nazko Cone is due to the complex geology of the subsurface. Nazko Cone has North America's only commercial deposit of black scoria and the most varied range of colours of any cinder deposit on the continent. Its carefully harvested slopes have supplied green products to the horticultural, masonry, and construction industries for twenty-four years.

An old lava source filling a crack in a jumble of older flows cuts up the slope of the Ilgachuz volcano. The darker stone lining the sides of this dyke are panes of volcanic glass.

When the glaciers melted away, this piece of breccia settled down so lightly that the ice seems to have melted around it about five minutes ago.

The cliffs at Bull Canyon formed when their lavas poured into the shore of an ancient river that filled this valley. Today, the Chilcotin River still flows in that old river's bed. For thousands of years, the pit house village on this site was a major meeting place of the Secwepemc and Tsilhqot'in people.

An underwater lava tube at the base of Bull Canyon is capped with pillow lavas, which formed in characteristic ovoid shapes in the currents and eddies of the ancient river. Pillow lavas are plentiful under the oceans but much rarer on land. The strength and direction of the ancient flows can be determined by the shape and alignment of the pillows.

The Chilko River leaves the mountains only a few kilometres from the sea, and races east to join the Fraser. On its long journey through the volcanoes, it lays a track for salmon to follow it back. Here it cuts the way down through the basalts and ash of Lava Canyon.

A Story of Patterned Ground

One of the most stunning experiences among the shield volcanoes of the West Chilcotin is to suddenly realize that many kilometres from the first hints of human activity you are completely surrounded by rocks and mountains that have been intricately sorted for size and colour. One mountain may be sorted down to grains of clay, the next to pebbles the size of chicken eggs, and the one beside it to ankle-breaking rocks the size of basketballs. This rubble near Pipe Organ Mountain was sorted for size and weight by the freezing and thawing of underground ice.

Lines and Circles of Stone

Exquisite patterned ground is ever-present in the Ilgachuz volcano, where gravity and ice often work together to form shapes like musical rivers (left), or can make the land well up underfoot in springs or stepping stones (above).

THREE-DIMENSIONAL SORTING

Depending on the volume of large stones mixed into fine-grained soils and the frequency of freeze-thaw cycles, patterned ground takes on a wide variety of geometrical shapes, like spiritual labyrinths. The rocks that make up these polygons in the Rainbow volcano decreases in size with depth.

Overleaf: Even long after the glaciers have left, the stone of the Rainbow volcano continues to flow down in their tracks.

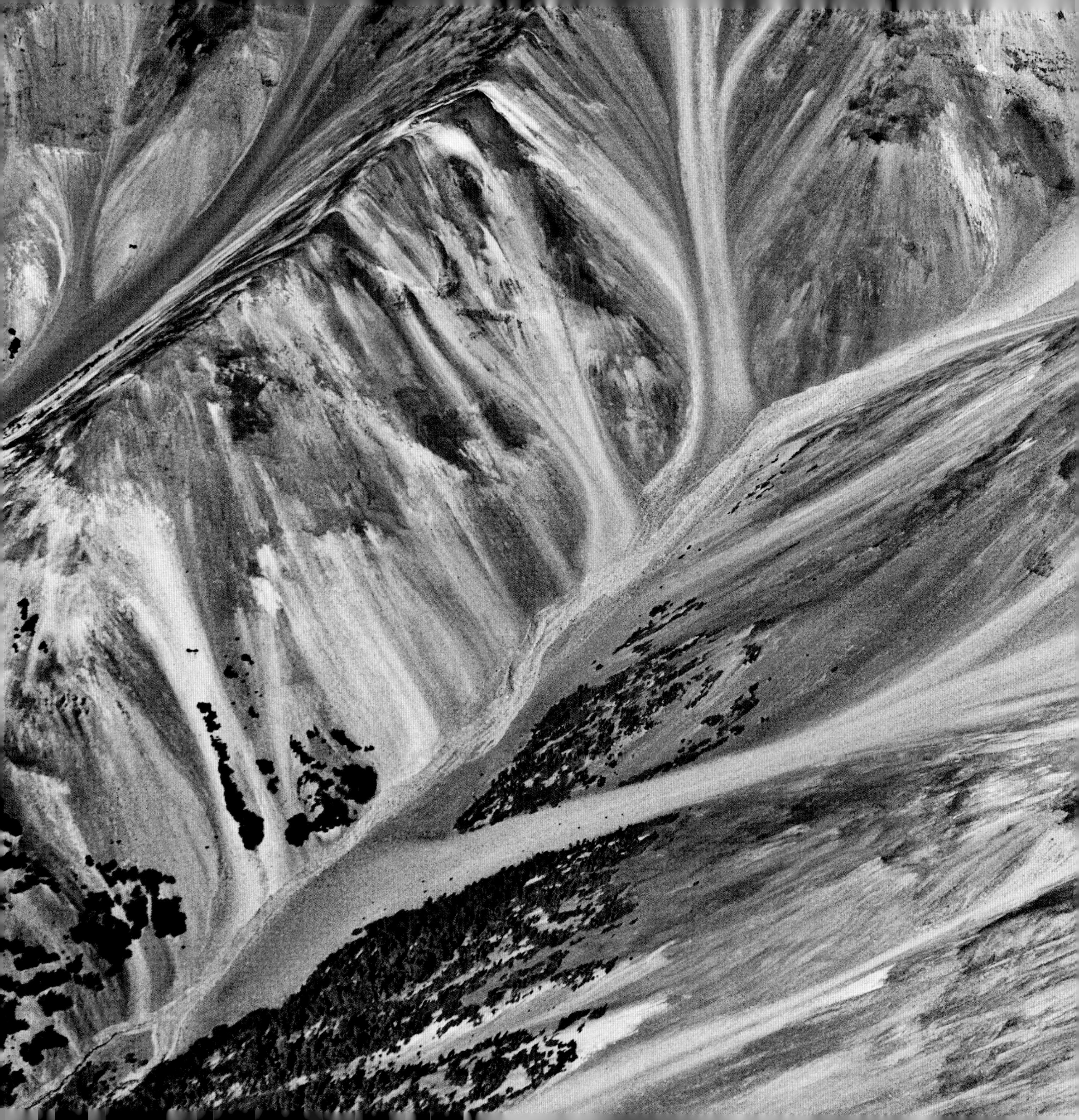

The Promise of Rivers

There is the Deadman, where the stone flowed like fire,
and black Vidette Lake, at the centre of the world,
among crumbling islands the moon blew in from the sea.
There is the Bonaparte, thin stream in red stone,
and Pavillion Lake, daughter of Mars.
There is the Fraser, stream of pithouses and smoke shacks,
draining the sap of the sky and the silver of the moon,
the rain turned into a ladder for salmon on the road of the stars.
There is the sea's throat gone to raw smoke.
There is Fish Lake and Big Creek with their loons and their reeds.
There is Cotton Lake in its bowl of sparrows and thatched grass,
and Beecher Lake, gravel cup in a shaman's sky,
and Hohium, where the thunder strikes, and Big Bar, where the glaciers died.
These are the waters that flow down from the Mother,
these are the ocean the clouds have lifted to the air.
There is the Mahood, where the cedars drink stone, and the Canim,
which catches the wind, and there is the Fraser's songline,
the San Jose, Lac La Hache, the Walker, the Bridge,
Horse Lake, Lac des Roches, and the Thompson's bright rapids,
in their bed of horsetails and herons against the wall of the world,
and there is the Fraser, river of candlefish and glass,
trail of polished stone and the sheep that follow the sun to it,
and their hunters who follow the wind back up to its floods.
These are the waters that flow through our hands now,
that pool in our lungs, and beat in the atria of our hearts,
and these are the waters that fill the hollows of our footsteps
in the dust of volcanoes where the earth split and spoke fire.
There is the Beef, in its valley of willows, and the Chilcotin, in its bed of salt.
There is Puntzi Lake and there are its pelicans and white waves.
There is the Dean, and its trout rising on evenings in August
to pluck the stars from a still untamed sky.
There is the Tusulko, Hump Creek, Blue Canyon Creek, and the Corkscrew:
organ music in the beds of the rivers of ice.

There is White Creek, Orange Creek, and the Ptarmigan Basin,
Shag Creek, Smoke Valley, Lunch Lake, and Goat Pass,
where the stone-footed climb to the end of stone,
where the waters that begin the return from the winter
first feel the winter return. There is Stum Lake, by Anahim's Meadows,
and Alixton Creek, where the stone crumbled.
There is Tatla and Nimpo and Anahim, where the stars spawn,
and there is the Fraser that carries the first silver people down to the sea.
These are the waters that flow through the Mother,
that carry her down, that we drink, that we honour.
There is the star road and the road of the waters.
There is the Canoe, and the Dog, where the land split in two,
and Alkali Lake, Swan Lake and Spring House, where men
gave the wind to horses, and they ran with it and will never return.
There is the Rose, and the Knife, and the Beaver,
and the Roe and the Latremouille. These are the waters that bring the sky to us.
This is where the trees come to the shore to drink stone.
There is Sulfurous Lake, Watch Lake, Sheridan and Green,
and the round eye of Sepa, and Tatton and Succour and Spring.
This is where we come home, where we hear Sen'klip, our ancestor, call,
and on still nights the high song of the aurora,
and the snow underfoot, and the fire and its summer.
These are the waters that bring us back, again and again.
These are the raindrops, the snowflakes, and storm.
These are the children of fire meeting the sea, and waves calling
and calling and calling and calling for shore.
This is the shore. These are the ways of the water.
This is the water in the stone. This is the stone within the song.

— *Harold Rhenisch*

Rivers of Ice

Three times, the Earth has grown cold above the Motherstone. Three times, winter snows have not melted. As the snowpacks deepened, their weight melded them into glaciers. During North America's Wisconsin Glaciation, 60,000 to 12,000 years ago, these solid rivers of ice eventually built up to a depth of several kilometres above the Motherstone.

The effects of these three ice ages were dramatic. Ice rivers carved valleys from the burnt cores of the hearts of volcanoes, and ground much of them into clay and rubble. Seas of ice called continental glaciers covered the volcanic peaks and all but the highest uplifted plutons of the Coast Range, then flowed over them and into the sea, past the edge of Vancouver Island. Most of the Motherstone's cinder cones were carried away by this ice; of the approximately 150 vents that spilled the Motherstone's lava only the hardest cores remain. As the glaciers groaned overhead, the Motherstone's pahoehoe, her top, ropy, glassy layers of lava, were ground to powder. Despite their size, even the shield volcanoes were transformed. The continental glaciers smoothed their summits, at times into broad plains. Localized glaciers followed the star-patterns of their streambeds and cut them into valleys. Rough contours were splintered, snapped, and sanded off, while in the Coast Mountains to the west the once smooth landscapes of uplifted country rock, volcanic plutons, stratovolcanoes, and ancient seabeds were transformed into jagged jungles of stone.

Despite all the activity flowing over, beneath and around her, however, the Motherstone has remained remarkably unchanged for perhaps 6,000,000 years. Three times the Earth warmed, and the glaciers slowed, then stilled, then melted where they stood, leaving their gravels in drumlins, eskers, morraines, and pebbled plains. Volcanoes have filled her valleys with ash, such as in the Bonaparte Valley west of Loon Lake and the Deadman Valley north of Skeetchestn, and

At the height of the Wisconsin Glaciation, only the highest peaks of the Coast Mountains were above the level of the glaciers. Such isolated peaks, called nunataks, were rare refuges of life in a sea of ice. This photograph shows the remnant of that glaciation, in the area around Mount Waddington.

In the Coast Mountains, glaciers continue to tear through the ancient granite plutons that once lay beneath a now-long-eroded volcanic arc. The top half of a glacial flow is hard and brittle, and breaks in deep crevasses under the stress of gravity. Beneath this layer, however, the ice flows as one solid, unbroken sheet. The grey bands on the glaciers are made out of rock that has fallen onto the ice from the valley walls.

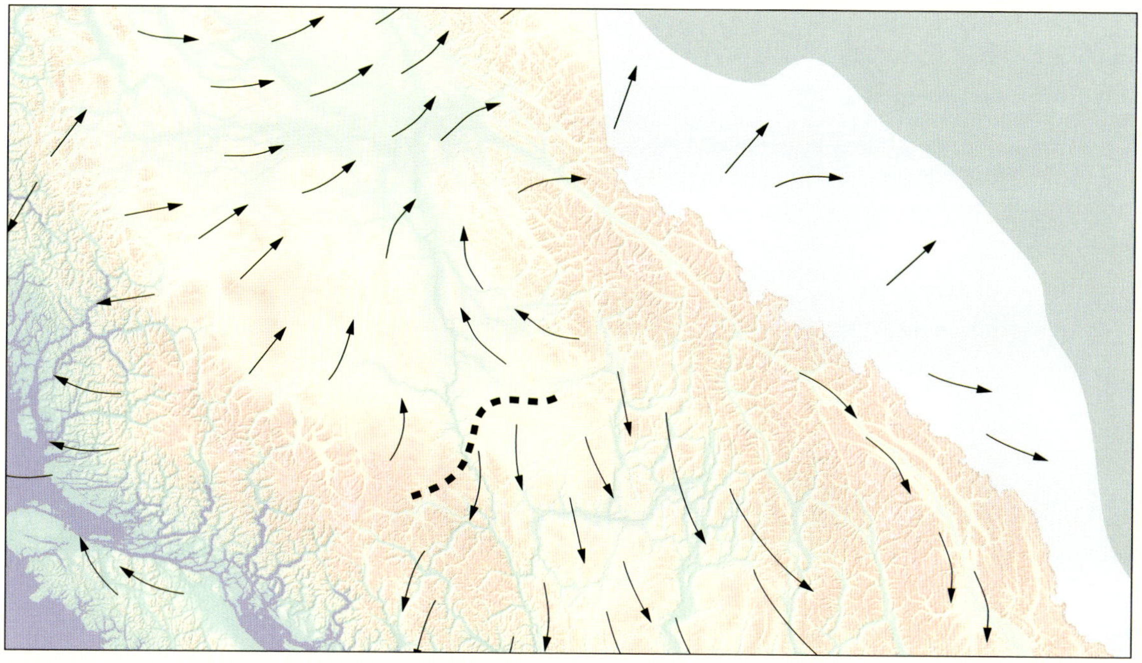

At the height of the Wisconsin Glaciation 14,000 years ago, British Columbia's glaciers divided along a continental ice divide (*dotted black line above*) that stretched from Big Bar to Bridge Lake in the South Cariboo. To the north, the ice flowed to the Arctic. To the south, it flowed through the Okanagan to the Columbia Basin. The ice north of the line lived on much longer than that to its south.

melting glaciers filled her rivers with silts and gravels, such as in Farwell Canyon on the Chilcotin River, yet each time her rivers largely refound their old channels. As the old seabed of the Explorer Plate forced the beds higher and higher, the rivers slowly cut down through the stone, creating the deep valleys of the Fraser, Thompson, Chilcotin, Bonaparte, and West Road. Because of their great depths and steep sides, these uplifted and downcut valleys appear to make the Plateau mountainous, but it is nothing of the kind. This is the land between the mountains. If anything, it is one giant mountain, slowly rising all at once.

The glaciers melted for the last time 9,000 years ago. Their meltwaters drained into ice-dammed lakes up to hundreds of kilometres long, with deep beds of alluvial gravels and silts. When the ice

Life returns in annual tides to the ice. Pink algae has colonized this ice field and the avalanches that spilled across it in winter and even by late July have not melted from its face. It uses these frozen waters to capture light.

Once all of the summer's 150,000 visitors have gone and the road is closed for the winter, British Columbia's most-visited cataract, Helmcken Falls, forms a dazzling, wintry volcano — in reverse.

As the Mahood River spills over Canim Falls, trickles that have eased down through cracks in the stone freeze in place. The simple pressure of winter freezing and spring thawing has been enough to break up the mountains and plateaus of the Motherstone into cinders and sand.

dams finally weakened and broke in the increasing heat of the sun, torrential rivers the widths and depths of the lakes themselves cut through the deposits — likely in a matter of hours. These massive flows are reflected in the vast, spreading, and often step-terraced valleys, hoodoos, and much smaller lakes and rivers that we live among today. Among a myriad of smaller waterways, they include the Chilko, West Road, Taseko, San Jose, Chilcotin, Fraser, Bonaparte, Quesnel, Horsefly, Nazko, San Jose, Mahood, Bridge, Williams Lake, Dog, Deadman, and Beef, all rushing, one way or another, down to the sea.

Although to the north of the Ilgachuz and Itcha shield volcanoes, where her flows were thin, the Motherstone remains now only as isolated whitecaps on waves of ancient rock, to the south and east she lies in a calm pool upon the land, reflecting the meeting of forces far underfoot that find balance here between the mountains. She is like an ice floe on the Fraser River in the winter, slowing turning in the current yet never moving downstream, although the river carries chunks of ice and snow inexorably past her and under her to the sea.

Among all of the Motherstone's rivers, the Fraser is perhaps the most consistent and yet the most changed. Although it still flows in its ancient bed along the Fraser Fault, it reversed direction 3,500,000 years ago. Before the formation of the Fraser Canyon, it flowed north, through a broad, shallow valley perhaps ten kilometres wide, and carried glacial flours and volcanic salts down the Mackenzie to the Beaufort Sea. During the Wisconsin Glaciation, the ice divide that separated the drainage basins of the Arctic and Pacific Oceans was along a line between Dog Creek and Bridge Lake, right over the Motherstone. To the north of the divide, the ice dragged gravels from older glaciers along with it, filling entire valleys and moving the continental divide 1,000 kilometres to the North, to Fort St. John. To the south, the ice flowed into the open Thompson Valley, with its ancient, land-locked salmon.

Basalt lava cools in huge columns at right angles to its cooling face. Wells Gray Park's White Horse Bluff erupted explosively into a subglacial lake 300,000 years ago, then melted a path up through the ice. It cooled rapidly against the ice wall, forming horizontal patterns.

When the glaciers were gone, the Fraser followed the land's new contours, sliced down along the fault line through the mountains to the south, and formed the Fraser Canyon. It carried the old land with it to the sea and laid it in a vast alluvial fan, called the Nitinat Fan, at the mouth of the Strait of Juan de Fuca off southern Vancouver Island. When people followed the river's trails north, the land became their story. They are our ancestors.

Others of the Motherstone's rivers are gone, of course, such as the ancient riverbeds at Barkerville and their gold, that lie now deep under gravel, or the ancient river that still seeps down around the Dog Creek Dome from Joe's Lake and Sting Lake, yet they are still there, and below them the Earth turns among the stars.

One drop at a time, rain and meltwater have polished the rough basalt of White Horse Bluff into the dry bed of a river. Since you can never scoop water out of it, its current really flows only in time.

A lava flow that poured down Dog Creek and Canoe Creek 3,500,000 years ago temporarily dammed the Fraser River and created a huge lake before the river broke through and flowed north once more. The dam was probably very short-lived. A pair of peregrine falcons nest there now, where the dam's remnants are uplifted high above the mid-Fraser canyon.

The Dog Creek cliffs are among the most spectacular of the entire Motherstone.
Columns like these typically fill the middle sections of exceptionally thick flows.

To walk on glacial ghosts like Table Top Mountain at the crown of the Itcha volcano is to feel the deep ice flowing, and to be that ice, even though it has transformed itself into grass.

This unnamed flow remnant on the Itcha volcano was turned into a mesa when the ice slid over it and simply carried everything but this central region away.

In Pinnacles Park, tall hoodoos of volcanic ash are preserved from weathering by small caps of stone at their crests.

A Young Volcanic Field

With an elevation of 1,440 metres, Kostal Cone is among the lowest volcanoes in Wells Gray Park. Although volcanic debris it blew out of its core still lies among the daughters of trees it would have blown down and set alight during its eruption, no charcoal has been found for carbon dating. An analysis of tree growth rings on the cone itself reveals that it last erupted only 400 years ago.

Two subglacial volcanoes, from two different periods of glaciation, lie along the Clearwater River near the north-south trending North Thompson Fault. Deep in the mantle below them lies the subducted boundary between the Juan de Fuca and Explorer Plates, which continue to divide beneath them even today. White Horse Bluff (*left*) erupted in a sub-glacial lake 300,000 years ago. Pyramid Mountain (*above*) erupted in a violent explosion through the ice of the Wisconsin Glaciation 9,000 years ago, but stopped short of breaking through into the air. Gravel from the glacier is mixed throughout the cinders of its cone.

On our planet, the soil takes on the form of the snow, the sun takes on the form of the soil, the snow takes on the form of the sun, and the result is flowers, trembling in the wind.

Every year, the yellow glacier lilies of Trophy Mountain in the mineral-rich volcanic uplands of Wells Gray Park put on one of the world's great wildflower displays.

In the willow thickets bordering the wet uplands of Trophy Mountain, white-crowned sparrows flit out over the flowers, and then back among the catkins, leaving their songs behind. Week by week in the meadows, new alpine flowers rise up through their sisters and claim the light. It is a wave that continues for months and then stills, as seedpods rattle their seeds in the wind and are brought down to the earth again by the snow.

Spirit Water

The water has an elemental energy when it is coming directly off a snowpack or bubbling through the rock. You experience it at its source. It is alive. We talked about this nearly every day when we drank from wild streams: the feeling of salvation as you came across this hot, baking rock, pulled off your pack, and the pure sensuality of plunging your hot self into cold, original water.

— Rita Giesbrecht

For millions of years, basalt flows born in fire have been drifting with gravity down to the sea. Here, the slow water of Bridge Creek helps the process along in downtown 100 Mile House.

Nearby, Moffat Creek Falls works the basalt loose near Horsefly. It has all the time in the world.

Finding Ancient Rivers

Where the Quesnel River shifts gravel,
the Cariboo River shifts gravel.

Between them, the false Solomon's seal and the Chinese grave-
yard,
the rusted stovepipes, and the water parsnips and thistles.

Men dug for dragon's gold here. They found it.
The rivers that shook them off in the rain

are not done yet with their workings.
We're not done yet either, not even here.

Sow and her cubs lined up on the road facing me,
teaching them what to look out for; teaching me.

— *Harold Rhenisch*

These sediments in Devil's Canyon were laid down on the bed of Glacial Lake Fraser at the end of the last ice age. Sister sediments to these form the bed of Georgia Strait between the Fraser River and the Gulf Islands, where they were carried by the catastrophic floods that emptied the lake around 11,000 years ago.

The natural colour of basalt lavas is a dull or ruddy brown. Exotic colours, such as these reds, purples, yellows, oranges, greens and blues in the Ilgachuz volcano are the result of mineralization from being cooked, sometimes for thousands of years, within the underground chambers of a volcanic complex. The green rocks in this image are coated with a thin layer of green volcanic clay. The orange rocks are actually white rocks covered with a thin deposit that has rusted in the air. The Rainbow Range gained its name from the brilliance of this thin and durable mineralized skin that covers the splintered, frost-wedged stones of entire ridgelines.

Rivers of Glass & Grease

The sea is still breaking against the shore. Among its waves, people live now, tiny and in wonder, in pain and in love, at work, at play, building homes and freeways, raising cattle and children, drilling for oil, setting up subdivisions on old branding corrals, drinking coffee, listening to songs on the radio, and talking and writing books and remembering. Together we have given this experience many names: North America, the Pacific Northwest, Canada, British Columbia, New Caledonia, the Chilcotin Back Arc, the Anahim Volcanic Belt, the Escarpment, the Intermontane Biome, the Wells-Gray Clearwater Volcanic Field, the Chilcotin Group, the Cariboo-Chilcotin-Coast, the Quesnel Lake Highlands, the Interior Plateau, the Gold Rush Trail, the Trail of Ranches and Rivers, the Heartland, and perhaps most simply, "the bush" and "the sticks." Its great arterial river carries the salts and ground stone of most of the Motherstone down to the Salish Sea, where orcas splash through it and salmon first smell their own land. We have inherited an ancient story.

Our ancestors came here on the trail of caribou and salmon. East of the Coast Mountains and north of the mouth of the Precipice, these nomadic hunters found the lone sentinel of Anahim Peak, where entire streambeds are pebbled with obsidian. Before the founding of the Old Kingdom in Egypt, Anahim Peak had become one of the centres of the world. For 4,000 years, its obsidian was traded extensively for thousands of kilometres to make cutting edges sharper than razors, many of which have been found at old fish cleaning sites in the eroded intertidal zones below old village encampments along the glacial fjords of the coast.

The price for obsidian was an oil made from the fermented bodies of oolichans, fish so oily that when dried they can be lit like candles. For the peoples of Western North America, who lived on

Black volcanic glass, such as this teardrop obsidian from a stream bed on Anahim Peak, fractures into the sharpest of all known cutting edges, almost at the atomic level. Today, obsidian blades are used for open-heart surgery. Incisions made by no other blade heal so quickly.

The historical trail between the Motherstone and the coast followed the track of glacial meltwater down through the Precipice Valley to Bella Coola. It is the lowest altitude passage through the Coast Mountains between Vancouver and Prince Rupert and served as a highway for 8,000 years. The edges of the Precipice (*left*), show the thinness of the plateau's basalt flows, as well as the ancient land beneath them. Just to the east of the Rainbow volcano, the hard plug of Anahim Peak's summit (*above*) protected its soft flows from the ice that carried all the rest of its flows north towards the Beaufort Sea. The story of the formation of Anahim Peak, and its pre-glacial shape remain a mystery.

A pictograph, drawn with a traditional mixture of red ochre and fish grease, is traced on a monolithic stone at Towdystan, at the eastern edge of the Precipice Trail. Many of the earliest stories in the world come from attempts to read our ancestors in rock. So often, the old stories are tales of monsters turned into stone so that the land would be safe for people. The monolith's best-preserved spirit figures were first photographed in 1923.

berries, roots, lean fish, and lean game, these oils were essential and became a culinary delicacy. Over time, the grease trade carried deep into the lands of the Walla Walla and Umatillo of the Columbia Basin, and across the Rocky Mountains to the Great Plains. In honour of this staple of diet and trade, the great river of the Motherstone was called the Ouregon River, the river of the oolichans.

In the 1700s, legends of the Ouregon told by the trading people of the Great Plains brought American explorers to the West, and the land that was soon called Oregon — a land of legend and myth beyond the edge of the world. The Motherstone is the heart of that country. When Alexander Mackenzie came overland from Canada in 1793, it was the Ouregon that he discovered, and one

of its major tributaries, the West Road, that he followed to Bella Coola and the Pacific Ocean, on the old Grease Trail.

On the trade routes of the Ouregon, grease and obsidian made their way south, and ochre and flint from cliffs in the older volcanic province of the Tulameen and Similkameen Valleys made their way north. By the time the first Europeans arrived in the folded mountain valleys of the British Columbia Interior, the Similkameen and Tsilhqot'in peoples were virtually identical, perhaps because control of trade from both ends made a lot of sense. The result was that for thousands of years, in that time when maps were made not with paper and compasses but with trails men created as they walked through the land, it was in the country of the Similkameen, in Princeton and Granite Creek on the western flank of the Okanagan Nation, that the Chilcotin ended, rather than along the silt cliffs of the Fraser River Canyonlands between Big Bar and Quesnel.

This is an old story. For 12,000 years humans have lived on the lower slopes and plateaus of the Motherstone, especially in the warm troughs of its river valleys and along its turquoise-blue lakes. They have followed the rivers to trade with the people of its sister stone, the Plateau People of the Columbia Basalt, and with the people of the fjords. They have lived where the salmon have come, and the deer, and the bears, yet despite this ancient history and its deep connections with the land they are still not the first people. Without question, that honour goes to the salmon. During the time of the glaciers, cut off from the sea in the cold, ancient waters of Glacial Lake Thompson, the salmon found refuge from the ice where ancient islands lay jumbled together like rocks in a glacial moraine. When the ice melted, the rivers flowed again — some of them even among and between those islands. The salmon flowed to the Salish Sea with the reborn rivers, and from there into the Pacific, and returned to spawn, again and again, and fertilized the land with their bodies. These rivers of salmon, streaming by their millions as tiny silver smolts down to the sea, and back

Packers brought sugar up through the Precipice to the Motherstone and Williams Lake from Hudson's Bay Company ships in the Bella Coola Valley — a trade that continued up to Williams Lake's modernization in the 1950s. They left only a few broken plywood sugar frames and rough-cut trails and blazes to mark their passage.

as strong, red muscles to the waters that still flow over ancient, oceanic rock, are one of the Earth's great stories of abundance.

As salmon populations increased 4,000 years ago and spread deeper and deeper into the river networks as the great herds of the grasslands that had come when the ice left the plateau above them dwindled, the people of the central and southern Motherstone settled down from their own wanderings and divided their year between winters in villages along the rivers deep below the plateau, and summers following the progression of plants through the grasslands above them, just as the caribou had likely done for 5,000 years before them. Now the grasslands are given to cattle, which make the same migrations to their summer ranges.

Little, though, has changed. The people are still in their villages. The caribou still migrate across the very northern edge of the grass, deep in the Ilgachuz volcano. Time still takes on a physical dimension here. It is still all present at once.

For 9,000 years, these patterns of weathered and frost-wedged rock in the Precipice have filled the stories of travellers. These stories of frost prying the cliff faces apart column by column can all be found within a fifteen-minute hike along the edge of the plateau. Put together, they form an ancient map of the land: one that is walked, not read.

The Ouregon River follows the Fraser Fault through the heart of the Motherstone. It is one of the great ironies of history that the land of the Columbia River came to be called Oregon, and the land of the Ouregon River came to be called British Columbia. No matter. The rivers still carry the mountains with them, as the remnants of the Juan de Fuca Plate deep below lift them ever closer to the sun.

The Ouregon River

Brothers, when they cut our country into two, they never divided us.
You got the Horse Heaven Hills and the Columbia Basin basalts,
streaked with yellow lichen and thin blue light;
we got mixed needle-and-thread and porcupine grass
and the basalt esplanades flanking the Dog Creek Dome.
You got the Youngs River Delta and called it Oregon.
We got the Ouregon River and called it the Fraser,
with its Sheep Creek and its Jackass Mountain and its Kanaka Bar.
None of us held onto the oolichan long enough for it to matter,
and the grease trails, well, those have become a matter for horses.
We kept the root of the San Jose River at Lac la Hache, yes,
but we fenced it so men with silver belt buckles could grow timothy.
You held the Columbia's grey mouth in the willows of Dismal Nitch
and the last hours of the desert, yet flooded it, right up to the tar-scented
baby's breath and Granny Smith apples at Fort Okanagan
and the willow-leaved peach trees at the mouth of the Kettle
so men with turquoise belt buckles could raise cattle in the coulees.
Small difference. We both lost the scent of wild apples
and the Beaver Valley and the clematis growing over the porch.
We both lost the rapids and the whine of mosquitoes
and the ability to find our way in the dark by starlight.
We are all the sons of our fathers. Why you got the heavy light of Soap Lake,
or the weightlessness of its water, I know no more
than why we got the patterns of big sage singing their way
down the alkaline hills into the salt lakes of 70 Mile House and Clinton.
A black bear that walks up the Ouregon to the obsidian,
aa-aa basalt and lava bombs of the Itcha Mountains, is stepping
across the line that divides us just the same as he does on his way back.
For him, North is just a line in the sand. He crosses it. Then he crosses South.
He pays no duty for the privilege of coming home to us or the greater
privilege of turning around in the month of wild berries picking
and coming home again. He has passed so many times now, this way and that,
that the line men have drawn in the sand has become blurred by his footsteps.

— *Harold Rhenisch*

In Devil's Canyon, the Douglas firs, rabbitbrush and sage of the Cariboo-Chilcotin Grasslands have rooted on the slopes of an island from the far Pacific that has collided with the shore. Such fantastic landscapes likely form the ancient earth everywhere beneath the Motherstone. Once thought to be hundreds of metres thick, the Motherstone is now believed to be approximately five metres thick over 65% of its area. Beneath it lies ancient rock like this.

THE FUTURE OF THE MOTHERSTONE

A small group of woodland caribou from the Itcha Ilgachuz herd migrate across a melting snowfield on the edge of a climax grassland of northern rough fescue, an arctic grass at the southern limit of its range. Chris Harris had a dream about these elusive animals. After he stopped fooling around with maps and followed the dream, he found them here in Ptarmigan Basin, right where the dream said they would be.

Rivers of Light & Story

Those of us who have found our way to the Motherstone have known the light and air within her for years. On moonless August nights, the stars are so bright here that we find our way for miles by their light alone, across the grass, along the rivers, and through the trees. The stone itself is solid, watery, and airy under our feet. We aren't actually standing on bedrock at all. Somewhere below the level of words the Earth and our dreams are one. We are standing in them. We have watched this dream Earth turn its shoulder to the moon as the Earth's shadow has swung across the lunar plains and craters and turned them to darkness, and have been blown this way and that by rivers of ocean cloud and storm. In a billion years, perhaps, this wave will have stilled. For the Earth, this great surge is nothing more than a wave breaking on a beach and in a few seconds taking away a man's or a woman's footsteps, as if he or she had never stood there.

Although this volcanic country has been a humanized landscape for 12,000 years, it is still scarcely known. It has been studied only superficially, by only a handful of volcanologists, each of whom has specialized in a completely different area. Government geologists have surveyed it, but mostly from its glacial valleys, where they have made annotations about formations and minerals before moving on. Industrial geologists have studied it in connection with mining claims, at points where the glaciers have scraped aside the basalts and revealed the older volcanic formations beneath them. Hunters have come from the cities each fall, to be led on pack trips over the most accessible passes, then into willow swamps by the ranches in the glacial valleys, where it's a lot easier to pack out a ton of moose meat than it is a few days further into the high country. Others have come for woodland caribou that pass rapidly through

AT HOME IN THE MOUNTAINS

We explored the Ilgachuz volcano from our base camp on Carnlick Creek, rising early and sometimes not returning until the late evening light. I learned there how beautiful sculptures are when they are made by nature over millions of years and never touched — natural rock rivers, and simple paths like a stone that's been overturned because a goat went through. It felt like eternity. One night, we even woke to snow and the chill of winter, but it was gone by noon.

— Mike Duffy

In mountains that make their own weather, the way deeper into the mountains is often made without bearings, in a whiteout, such as this hike through Misery Pass in a summer snowstorm (*above*).

Standing in a circle of frost-turned stones, Chris takes his bearing on magnetic north, looking for Goat Pass and the trail further into the Motherstone (*right*).

The paths of the mind are the real trail into the mountains. When the weather clears, the way of the map and compass vanishes, replaced by the eyes. The mountains take shape as eternal living beings, scarcely separate from ourselves (*facing page*).

The Motherstone sits in stillness, while the turbulent forces of the Earth flow past.

the glacial valleys in the early snows. Few packers or hunters have ever walked the highest volcanic slopes. They are too high and cold. There is little to hunt, and what there is is well hidden. There has not been enough time since the glaciers to warm the land and make its extremes worth the physical cost of leaving more sheltered, warmer terrain. Our fellow traveler, Mike Duffy, relates how he was bathing in a pool of meltwater in the Carnlick Creek krumholz, when a caribou came right up to inspect him closely. It was only when Mike stood, 6'4" of white, naked flesh, arms awkwardly at his sides, and streaming with water, that it bolted at such a ghostly apparition. To tell the story of the Motherstone is to tell a story of an Earth men and women and children have not known for thousands of years — a profoundly non-human Earth that becomes human as we walk it. Just as the Secwepemc became a people when Coyote tamed the land, and the Germanic and Nordic peoples of Europe found their names when they moved into the great forests of the central European plains and the fjord-lands to their north, we have become a people now, as we walk on the Motherstone together. However it has been told, and whichever land it has been told with, our first Earth is a universal story of a common ecological heritage.

It is a story that continues to be told from the ground up. After a century and a half of colonial culture in British Columbia, the new cities of the southwest have grown into vast metropolises of concrete and freeways, symphonies, film industries, huge bridge-spans, and rapid-transit overhead trains. The Motherstone remains sparsely populated and little known. None of her traditional colonial industries — ranching, hunting, logging, and mining — thrive as they did even twenty years ago. In broad social and political terms, she is to the side of the road or along a road to somewhere else. Highways climb onto her escarpment, cut across her, lead through a few sparse towns with their gas stations and mechanized mills, and then drop off her on steep grades, but they do not lead

Autumn Journey

The Cariboo-Chilcotin region is one of the Earth's great high altitude landscapes. Lifted in the west to an altitude of 1,100 metres by the formation of the Coast Mountains, it slopes down to 400 metres along the Fraser River, before rising again to Williams Lake and 100 Mile House. Here at an altitude of 1,800 metres on the glacier-carved tablelands of the Itcha volcano, a packhorse ride can feel like a trip into space itself. When I told my old friend, mountain man Bill Mero, about my trip onto the Itcha volcano from its southern flanks he invited me to come back in the fall, and to approach it with him from the north. He had, he told me, a surprise.

— Chris Harris

As Bill and Chris crossed the highlands of Table Top Mountain, Mount Fukawi rose like a hunting lion or a ruined giant out of the body of the Itcha volcano. They passed it by. As they passed south through the low angles of the fall sun, the light changed more quickly than the pace of their horses. Because this almost arctic alpine landscape lies within the rain shadow of the Coast Mountains, it shares many characteristics of the low altitude grasslands along the Chilcotin River.

The explosions that formed this post-glacial vent on the Itcha volcano are not the end of its story, but its beginning. In the fall, exquisite patterns of life weave across it, like a glow coming off the stone. This still-living vent was Bill's gift to Chris.

into herself. Gravel roads do, a little, and industrial forest access roads, but even they quickly peter out into potholes and boulders and scarce room to park a truck among the ripped-out mufflers and shotgun shells. The truth is, if you're going to find the Motherstone, you'll have to get out, off the road, and walk, step by step by step. A volcano that looks beautiful from a distance is quite another thing when its ten- or twenty-kilometre-long flanks of crumbling cinders are walked along trails that goats have followed for 10,000 years, when those goats are your only companions and civilization is a train of packhorses and a wrangler who will pick you up at his camp over the hill in exactly ten days time, when 'over the hill' means 'over a mountain range and somewhere in a valley twenty kilometres long,' and there will not be another pack train in until next summer. The world of ideas simply falls away. After that, you have to find your way back, and find the words and political and social forms to express the moment when you walked right into the land and only the land returned.

The singers of the ancient migratory rivers of lightly overturned stones and silt baths are the migratory and stationary herds of California Bighorn Sheep of the Fraser, Chasm, and Chilcotin canyonlands; the caribou of the Itcha Ilgachuz Herd, feeding on an untouched, unmapped arctic grassland; and the mountain goats of Bull Canyon and the high, crumbling slopes at the cores of the big shield volcanoes. They are the last of the great herds that brought humans to the high plains of grass that sprouted in the wake of the glaciers.

Those herds are not, however, an ancient history; when Europeans arrived in the 1820s, caribou were still present in 100 Mile House. Throughout the Motherstone now, they have largely been replaced by domesticated herds, the moose that came back to the land in 1923 and move like highway graders through the brush, and the wild horses of the Esketemc people — rodeo broncos and cattle horses with the wind in their eyes and the scent of sagebrush in their

In a world of 7,000,000,000 people, in which world governments are pouring huge sums of money into the exciting stories of the exploration of Mars and discovery of planets outside our solar system — perhaps even as a future home for humankind — much of the Motherstone still remains untouched. The planet it speaks of is one few of us know anymore, yet which all of us are designed to thrive in. *Ilgachuz volcano.*

The Chilcotin River's 241-kilometre-long journey across the Motherstone begins in the un-named streams on both sides of this monumental plug in the Itcha volcano.

As I've become more acute as an observer of light, I've learned to approach my journeys with an open mind. I've become more intuitive and I know better when to push the shutter. I see, and track shapes. The shapes create emotions. I follow them.

— Chris Harris

manes, that spend long, dusty summers in the Cache Creek Terrane.

The social will to preserve the richness of the plateau is as ancient as the land sense of the first peoples, who knew this land as the spirit of their ancestors, and preserved it to preserve themselves. It is also as new as the Itcha Ilgachuz Provincial Park, formed to protect the woodland caribou in their last refuge, and the Junction Sheep Range, the grassland triangle above the confluence of the Chilcotin and Fraser Rivers, formed to protect the mother herd of North America's surviving California Bighorn sheep. Both of these far-seeing actions were made in 1995. They stand on a foundation of science, deep vision, and commitment. They are a model for us all. Our Earth and the species who are our brothers and sisters don't need protection from glaciers but from the stories we tell that do not include them.

We have come to it along many paths. We have come as teachers, adventure guides, mothers, fathers, engineers, scientists, and poets, but we have all found our way to the black and red rock of this volcanic sea. Even so, it is not our story. No man or woman can own it. The Motherstone belongs to itself, and we to her, and through her to each other. If we do not continue to protect the Earth she has preserved for us in its purest, most essential form — the Earth that is the cutting edge of our survival in the coming climactic and social upheavals that may very well await us — we will not have a story to preserve us. We will not have a mother stone. Without the hope gained from extending the visionary physical and social legacy of her protection, we will be lost among the rocks and will have only a wilderness as our mirror. It will be a mirror of our diminished selves.

This volcanic dyke in the Itcha volcano was formed from lava pushed up from a crack in a previous flow of softer stone. Later, the glaciers that sculpted the valleys around it stripped it of its mantle, and even shortened it, but proved unable in the end to bring it down. Even 9,000 years of weather have not succeeded yet where they failed. It stands in balance now, as it always has, evoking other walls — first of stone, then of ice — that fill the air around it like ghosts and hold it up, even today.

There is art and order in the Itcha volcano that shows no sign of the presence of our species on this Earth — likely not even when Chris made this image. There was light, certainly, and stone, and air, and there was one sound: the sound of an animal's breathing and the creaking of its equipment as it set up a tripod and looked through a viewfinder and became, for an instant, human, before it turned and walked into the Earth once more.

The materials of a work of art do not disappear in its making but shine forth through it as if for the first time. Here, the hard cores of the vents in the Itcha volcano rise out of the plateau like islands out of the sea.

At the Top of the Pass

A very difficult climb up a steep, waterless draw of crumbling stone brought us to the most powerful moment of our ten days in the Itcha volcano. When I finally got out of the rocks and crested this pass, instead of open country I found this elemental landscape shrouded in smoke from a forest fire in the Yukon, far to the north. In my bones and my heart, I felt that I was watching the actual eruption of the volcano. It took me many minutes before I could compose myself enough to start shooting.

— Chris Harris

The Star Road

Three hours lost on unmarked roads across the plateau,
government map sketched to a principle other than cartography,
roads intersecting with roads miles from their locations in dust,
lakes thrown into entirely different watersheds
as if the map had been drawn up by Vasco de Gama,
by memory, freehand, without a sextant —

I find my way off the plateau by the Marble Mountains,
that I have only ever seen from the other side,
not even paying attention to the map anymore,
or the government, or our destination, Dog Creek,
which the Ministry of Forests put miles out of place,
perhaps because there are no forests there, who knows;
just following the lay of the land within me now,
just following its folds, just making dust.

New car, kids hungry in the back seat, road turned to deep ruts
of Cariboo mud turned to concrete that some 4x4
cowboy splashed out in the spring, car filled with tension,
gas gauge going down, all of us thinking
of being lost here and never getting back home,
nobody wanting to say it, Sunday getting very old.
Three metres from the Pavillion Road the gate is locked,
for inexplicable purpose, ruts continuing underneath it,
and a chill goes down my neck
at how the Ministry of Forests led us to this. The turning

around in the rudiments of a branding corral,
the following of the ruts again, this time without hope,
this time with silence as thick as moonlight,
this time with the mountains behind me;
finally taking my family out along the powerline
to Beaverdam Lake and Chasm, and then gas at 70 Mile,

and heaping the kids with candy and chips from the store,
and then driving north again, on the black road.
Half a million kilometres — twelve years — later
the highway is blocked at midnight, above Clinton
and its old law court and wagon roads on the way to Pavillion.
The lights of Chasm shine through the black trees,
unreachable, along the curve of stone. Six cars up,
a life is scattered across the pavement. Pickup on its side,
pools of gasoline, blood of the beast, police stopping traffic,
blue and white carnival light show, long lines of cars
no longer heading north or south, people milling in silence,
the 3 a.m. drive down Coyote's Road to Vancouver,
looking for a new home for my daughter (and no luck),
then five hours back, up the escarpment,
doing the salmon dance, in the old car crumbling to rust,
for a four-hour wait, chained to the road,

and then it comes back, how I got lost on my way home once,
how I never found Dog Creek, only grouse hunters at Gustafson,
only bluebirds spilling out of the poplar trunks
and the tracks of someone who'd gone out too early in the spring,
so I pull out of the line of traffic going north
and head west in the dark, my headlights cutting a tiny path
under a massive dome of night. I could have done it with my eyes closed.
I could have done it with one hand tied behind my back.
I did it with my daughter beside me,

just like those years in the bed of a southern volcano,
when I used to pull her through town in a wagon and she would squeal,
before we bought the white car to move to the plateau,
when there were still forests. A few trucks parked out in a clearcut.
Man flags me down. Asks me if I know the way.
Around us a starlit clearing of replanted pine.
Above us all the time in the world.
I assure him I do, send him off on my dust road to the east,
then bring my daughter home by the stars.

— *Harold Rhenisch*

The Zen Garden

From our tents on the Carnlick Creek, we saw light reflecting off the mountain. It was so bright it drew us in. As we hiked towards it the next day through the scrub, it kept getting brighter and brighter. There was broken, low cloud, and shadows would hit the stone as the sun moved through them. Then I saw my own shadow for the first time there, like a shadow on a beach of white sand, and realized that before then the only thing that had ever obliterated or dimmed the light was cloud, but now, all of sudden, it was my own human shadow. We called it the Zen Garden.

Wherever we hiked for the next ten days, we saw the Zen Garden shining out at us. There was no way we could leave it and not return, so on our last day we hacked through the bush again. When we sat down on the edge of the garden to get out of the wind, the light was buzzing off the rock. I felt like I did when I first saw the Cariboo. It felt like home. I can't get it out of my head. I was one with everything. Part of the landscape was an extension of me, and I was an extension of it. When it snowed on the Zen, the colour was unbelievable. It was the first thing that melted off, so you could see the light there. It was the hardest thing to leave when we left the Ilgachuz.

— Mike Duffy

A Woolly Story

Often we would be in the krumholz, those bunches of tangled firs, to have something to eat or drink out of the wind, and find that the goats had bedded down there before us. They were unerring in their ability to get out of the wind and find little sheltered spots. They knew every nook and cranny; the whole slope would be an open, unsheltered space, with one little nook where there was no wind, and they would find it.

— Rita Giesbrecht

At the point where earth, rock, and air meet, light springs to life. On the middle heights of the Itcha volcano, Chris found these goats grazing on grass turning sunlight into sweet sap and sunning themselves on sun-warmed sand. They truly are children of the sun.

Rivers of Beauty

Beauty is an elusive quarry. So are mountain caribou. To track large game like that, a man might go into the mountains with a flint and a blanket, often without telling his family he was even going there, and he might not return for months. To achieve that sense of place you have to stop thinking that you are going deep into a wilderness, because you are not. You are just going deep.

Here in the west at the edge of the West, after having been kept alive for generations by backpackers, fly-fishing guides, bush pilots, big game hunters, and a few aboriginal families, Beauty's story can finally be told. The land is in our bones now. The Beauty of the first European explorers was the fear of a tiny human being before the infinite, raw immensity and majesty of God. Today, that sense of infinitesimal size remains, yet in the face of global warming and species extinction the sense of insignificance has vanished. How we, ourselves, look at the world matters now.

Beauty is not ours alone. Mountain sheep seek out cliffs of a certain steepness for their refuges because in landscapes perfectly matched to their bodies they are complete; caribou seek out landscapes of a certain coldness and remoteness; porcupine grass grows only in certain soils with certain amounts of precipitation; and humans respond with serene joy to high cliffs looking over water to spreading green trees and to views of distance and light. As humans, we call beautiful these environments in which our earliest African ancestors most thrived. When we respond with art, emotion, or deep thought and contemplation to images of this kind, it is these ancestors who are leading us to them. It is to them we give voice when we make such images. We see with their physical eyes, eyes that are millions of years old and that have been with us before we were conscious of what they saw.

Our ancestors include the Europeans who brought us the language, science,

After a three day bushwhack through almost impassable swamps of willows and mosquitoes, Mike Duffy and Rita Giesbrecht pick their way carefully across the crumbling core of the Rainbow volcano. Each footstep is an extension of the mind. It feels its way through the rock.

In the joined flows of debris and sunlight down the Rainbow volcano, beauty forms when two complementary curves follow a line through space. When they are laid around three-dimensional shapes, they create physical snapshots of time, like these images made separately by two photographers: the volcano and Chris.

Chris was tracking the intense colours of the Rainbow volcano after a squall, when he found this grass using rainwater to sketch the finest of lines between two debris flows.

A mountain does not stand in light; it creates its own light. It does not stand in space; it creates its own. It changes with every particle of light that strikes its stones.

and religious metaphors that give our story of Beauty familiar form: the photograph, the awe-inspiring contemplation of God, and images of Paradise. Our ancestors, however, also include the Tsilhqot'in, Secwepemc, and Carrier, who gave us images of sacred journeys and the unity of body and soul, as well as the animals and fish of this land, the forms of the land itself, and the way the light lies across the land as it does nowhere else. To walk in the Motherstone today is to be present when this land sees itself for the first time — which is a moment that does not necessarily ever come to an end. It is to see the living Earth in her ongoing eruption and to recognize her as a mother.

There is no other planet like this. Our bones are made of this rock, our flesh is made of this soil, our eyes are made of this starlight, our words and images and ideas are made of this soil, and though this vision may have been largely obscured by the social tensions, pressures and catastrophes of history, a place remains for us to find our way home again: the Motherstone. Clear your eyes here. With new eyes and new surety in your footsteps, find the Earth everywhere you look. Go, and look, and wonder. When you meet other travellers, tell them what you have seen. Tell them in old words and old ways and tell them in new ones. Tell them what they already know in their bones.

Tell them what they tell you.

Tell them they are home.

As a photographer, Chris can talk enthusiastically for hours about focal lengths, wavelengths, pixels and Pantone colour calibrations. All the time, the silicate grains in this basalt take in photons that have travelled 149,000,000 kilometres from the sun in eight minutes and give them back minutely diffracted, in photographs that develop within the body of the stone itself. By the trillions. A thousand images of this moment, from a thousand different points of view, would not be sufficient to exhaust its potential.

The art of the sun. *Rainbow volcano*.

Overleaf: Watching time hold itself in place.

A hike through the Rainbow volcano means a thirty-kilometre-long trek across slopes of rocks the size of plums, all burnt into rich colour by fire. The heart of the volcano is the orange section in the centre right of the image.

Horse outfitters David and Joyce Dorsey live on the southeastern slopes of the Rainbow volcano and consider it their home. Here their pack train moves through the volcano's southern cinders (*above*) as it begins its descent into the long U-shaped valley the glaciers cut through the volcano's heart (*right*).

BEING PRESENT IN THE LIGHT

In photography, the gift of light provides a vast array of shapes, lines, textures and tonal contrasts. Photographers arrange these elements in ways of deep resonance. But when a photographer finds himself in the middle of shapes already perfectly balanced, what then? On Chris's last day in the Itcha volcano, he just stood there at first, in awe, then slowly began to exchange a few words with Rita and Mike. As the moment unfolded further, it was as if they were talking to themselves out of the land. Chris carefully picked up his camera then, and moved into the relationship with the land that the moment's mood had given them. For the rest of the day the three travellers moved through that visual expression together, entranced. Mike Duffy calls this one day the experience of a lifetime.

As water-saturated oceanic plates are subjected to extreme pressures deep underground, the water they release helps melt the mantle above them. From this magmatic source, and the lava's interaction with water and air as it breaches the Earth's surface, come the diversity of patterns and colour that we walk on and turn over in our fingers, and feel.

Walking on Naked Ground

I have always felt that rock is very alive. I'm too pragmatic for visions. I feel — I know — that my cells are made out of the same stuff as the dirt. The forms of the mountains are the forms I see in my own body and that of the voluptuous thighs and bellies, the breasts and vulvas, faces and moods of all women — not in some metaphysical way but because I'm already there. I simply reconnect. I am where I belong. I am completely an earth person. I don't like being on water. I don't like being in air. I don't like having my feet too far from the ground, and I don't like wearing shoes. I experience the planet by walking it. In the volcanoes, I was attracted to details — most often to plants, flowers, and berries and little water pools and streams. If there's no vegetation or dirt of any kind on a rock fall, you know that there's a good chance it will move under your feet. You can't just walk in those landscapes. You have to determine whether or not a rock will take your weight, before you put your foot onto it. If you can't take a step and find some stability, it's very possible that the slope will start moving and take you with it.

—Rita Giesbrecht

This glimpse of Pipe Organ Mountain through the light that bathes it is seen best from a point deep within your chest — from your bones. Everything else is the light dancing.

Chris noticed these glacially-ground blobs of lava and precipitated minerals in the Rainbow volcano claiming him, and managed to catch the moment through a tiny airplane window.

Overleaf: A photographer can't actually make this photograph of an ancient dyke in the Rainbow volcano — he can only frame it at the point where his human attention begins, and ends. Life, too, is about its frame.

A herd of mountain goats crosses a ridge line on the flanks of Pipe Organ Mountain towards the capstone spur on the far left. Like us, they come from the rocks of this planet. Like us, they are the land walking.

Overleaf: As the land is raised into the sky by the subducted oceanic plates beneath it, it grows increasingly exposed and begins to erode away into dust. The catastrophic floods at the end of the Wisconsin Glaciation greatly accelerated the erosion here in the Painted Chasm north of Clinton.

Timeline

Years Before Present

4,000,000,000 – 3,800,000,000 *Life forms simpler than bacteria evolve*

1,500,000,000 – 700,000,000 *Belt and Windermere Supergroup of sedimentary rocks deposited in eastern BC and eastern Washington State*

1,000,000,000 *Supercontinent Rodinia forms*

750,000,000 *Rodinia begins to fragment. Australia and Asia break off from eastern Washington and BC*

750,000,000 – 400,000,000 *Western shore of North America near Spokane/Nelson. A geologically quiet area*

670,000,000 – 600,000,000 *Local glaciation*

540,000,000 *Appearance of multi-cellular fossils*

530,000,000 *Fish appear*

425,000,000 *Plants invade land*

363,000,000 *Amphibians arrive*

370,000,000 *Oceanic plate begins to subduct eastward under continental margin*

350,000,000 *Reptiles arrive*

350,000,000 – 300,000,000 *Volcanic islands group together off the northwest Pacific coast*

370,000,000 – 280,000,000 *Volcanic arcs form along western edge of North America*

300,000,000 – 250,000,000 *North America and other continental fragments come together as the supercontinent Pangaea. Major climatic changes result*

250,000,000 *Mass extinction. An estimated 80 – 96% of species are lost*

205,000,000 *Pangaea begins to break part to form the Atlantic Ocean. North American Plate drifts westward. Volcanoes form in Stikinia and Quesnellia terranes in Intermontane Belt offshore*

185,000,000 *Intermontane Belt collides with and accretes to Omineca Belt*

170,000,000 *Coast Belt collides with Intermontane Belt. Volcanoes form in Omineca Belt*

130,000,000 *Volcanic islands of the Insular Belt approach the coast*

115,000,000 *Insular Belt collides with the coast*

170,000,000 – 85,000,000 *Magmatic arc forms in Coast Belt due to eastward subduction of Farallon Plate*

85,000,000 *Kula Plate splits from Farallon Plate and begins subducting beneath western North America*

85,000,000 – 50,000,000 *Arc volcanoes in Coast Belt erupt due to Kula Plate subduction*

65,000,000 *Extinction of dinosaurs, ammonites, and other species. Around 50% of total species lost*

50,000,000 *Coast Belt arc volcanoes stop forming*

55,000,000 – 40,000,000 *Deposit of continental sediments on Washington coast*

55,000,000 – 40,000,000 *Southwestern BC west of Fraser-Straight Creek Fault shifts 110 kilometres northward*

40,000,000 *Main movement on Fraser-Straight Creek Fault stops*

37,000,000 – 7,000,000 *Cascade Arc forms, south to north, along edge of subducting Juan de Fuca (remnant of Kula-Farallon) Plate*

31,000,000 *Beginning of minor Chilcotin basalt eruptions*

20,000,000 –15,000,000 *Olympic Peninsula lifts out of the sea*
17,000,000 –13,000,000 *Columbia River Basalt Group flows*
17,000,000 –13,000,000 *Yellowstone Hot Spot moves under the Columbia River Basalt Group and then eastward*
13,000,000 *Dike swarms in Bella Bella*
12,000,000 *Dike swarms in Gale Passage*
10,000,000 – 6,000,000 *Peak of Chilcotin Group basalt volcanism*
8,700,000 – 6,700,000 *Rainbow Range erupts*
7,000,000 – 5,000,000 *Modern Cascade Range forms*
7,000,000 – 5,000,000 *Explorer Plate breaks off from Juan de Fuca Plate*
6,000,000 – 4,000,000 *Ilgachuz Range erupts*
5,000,000 – 4,000,000 *Cascades and Columbia Basalt Group uplifted and eroded*
4,400,000 *Some of humankind's oldest ancestors roam around in Africa*
3,500,000 *Wells Gray-Clearwater Volcanic Field begins to form*
3,100,000 *Lava dam across Fraser River at Dog Creek. River flows north*
3,000,000 –2,000,000 *Second peak of basalt flows on Chilcotin Plateau*
3,000,000 – 80,000 *Itcha Range erupts*
3,000,000 *Beginning of Clearwater eruptions*
2,500,000 *Pleistocene Epoch Ice Age*
2,500,000 *Some of the earliest examples of possible hominid tool use*
2,500,000 – 7,600 *Volcanism in Wells Gray, along Nootka Fault extension*
1,170,000 *Second lava dam across Fraser at Dog Creek breached. Fraser River begins to flow south*
1,600,000 – 10,000 *Minor basalt flows throughout southern interior of BC*
1, 500,000 *Possible oldest evidence of use of fire by* Homo erectus
195,000 *Some of the oldest* Homo sapiens *fossils*
80,000 – 10,000 *Wisconsin Glaciation*
80,000 *Formation of the major peaks of the Itcha Range*
29,000 *Fraser Glaciation (of the Wisconsin Glaciation) begins*
25,000 *Ice advance from mountains into valleys and fjords*
16,000 –14,000 *Glacial maximum*
13,000 – 9,000 *Late glacial phase, glaciers retreated to early glacial positions (present positions)*
12,600 – 9,100 *Windust Phase of human settlement*
12,000 *Earliest human settlement on Haida Gwaii*
12,000 *Rapid uplift of eastern Vancouver Island due to deglaciation*
10,000 – 9,000 *Settlement of Chilcotin Plateau begins*

10,000 *Millbank Sound cinder cones in line with Anahim Hot Spot trail*

9100 – 6300 *Cascade Phase of human settlement*

7700 *Mount Mazama erupts*

7600 *Dragon Cone (Wells Gray) erupts; formation of Clearwater Lake*

7600 *First eruptions of Kostal Cone (Wells Gray)*

7200 *Nazko Cone eruption*

4500 *Late Period of human settlement*

3000 – Present *Major earthquakes on Cascadia Subduction Zone caused by subduction of Juan de Fuca Plate beneath North American Plate. San Andreas Fault earthquakes in California caused by slippage of Pacific Plate northwards relative to North American Plate*

2000 *Replacement of obsidian use on BC central coast by tool-grinding technologies. Reasons unknown*

1000 *Most recent eruption of Silverthrone Mountain (most northerly peak of Garibaldi-Cascades Arc)*

700 *Most recent eruption of Mount Edziza in northwestern BC*

500 *Big earthquake buries Ozette on Washington coast*

400 *Kostal Cone (Wells Gray)*

300 *(January 26, 1700, 9 p.m.) Cascadia Earthquake*

285 – 260 *Tseax Cone Eruption (Nass River)*

224 *Mount Shasta erupts (Cascade Arc)*

217 *Alexander Mackenzie follows the Grease Trail to the Pacific Ocean*

202 *Simon Fraser explores the Ouregon River and names the Thompson*

169 – 151 *American settlers head towards the Ouregon River*

153 *Mount Saint Helens erupts*

152 *Cariboo Gold Rush in a paleovalley east of Quesnel*

148 *Cariboo Road constructed*

146 *Chilcotin War*

130 *Mount Baker erupts*

116 *Mount Rainier erupts*

112 *Ruby Mountain erupts (Stikine)*

104 *San Francisco Earthquake*

90 *Reappearance of moose in the Chilcotin*

47 *Road pushed through to Bella Coola*

37 *Earthquake swarms under Mica Dam*

30 *Mount Saint Helens erupts*

21 *San Francisco Earthquake*

15 *Formation of Itcha Ilgachuz Provincial Park*

3 Oct 10, 2007 *Nazko earthquake swarms*

2 July 2008 *Photography begins for Motherstone*

Today: *You are reading this. Juan de Fuca-Explorer Plate still subducting beneath southern BC. Mount Maquinna mud volcano erupting undersea along Nootka Fault, 16 – 18 kilometres west of Vancouver Island (west of Campbell River). Geologists are continually refining this timescale in light of new evidence*

Future: *Additional volcanic eruptions and large earthquakes may occur in BC within your lifetime*

A herd of California Bighorn Sheep are at home at the very base of the Motherstone in the Painted Chasm. The multiple lava flows that formed the Chilcotin Plateau are at their thickest here, each one separated by a layer of ancient soil baked red by the 1200°C temperatures of the basalt.

Overleaf: Of the approximately 150 lava vents of the Cariboo-Chilcotin, few were hard enough to withstand the glaciers. One of these is Lone Butte, which rises above the ranching town named in its honour.

Reflections

Recently, I received a letter from a photographer in northern Germany who had visited my gallery in 2006. He described how he had travelled around Germany making images of 800-year-old cathedral architecture. I envied him. At his fingertips, he'd had spiritual grandeur that hundreds of thousands of people had admired for centuries. He had a direct connection to beauty. Only later did I realize that my work among the volcanoes was the same. I too had been walking in reverence through cathedrals, except mine were the cathedrals of the natural world. They were designed by the great architectural forces of the Earth: gravity, air, wind, sunlight, water, snow, heat, and cold. The only real difference between the cathedrals of Germany and mine was that hardly anyone had ever come to worship in the Motherstone. Over 20,000,000 years of beauty had yet to be discovered, and I was walking through it. I felt — and still feel — deeply privileged.

For two years I walked through these amazing galleries: the Itcha Gallery, the Ilgachuz Gallery, the Wells Grey Gallery, and the Bull Canyon Gallery. If you ever go to them yourself, remember that a photographer told you one day: walk; walk through every pass and hallway, through every valley and room; and walk slowly; and walk often, because each gallery is being re-hung every day.

I have my own guide. His name is John O'Donohue. Here are his words about vision: "Light is the greatest unnoticed force of transfiguration in the world; it literally alters everything it touches and through colour dresses nature to delight, befriend, inspire and shelter us. The miracle of colour is a testament to the diverse, precise, and ever surprising beauty of the primal imagination." In that light is the genesis of this book.

Words are vital to this book. Among the many things I learned from my previous book, *Spirit in the Grass*,

Like any art form, the mastery of the art of time is realized in a moment's perfect gesture.

is that scientists and I viewed the grasslands very differently, although with equal passion. I learned that success at bringing about awareness, knowledge, and a sense of respect for our planet requires that science and art bolster each other in a way that will bring about new forms for appreciation and understanding. When Harold Rhenisch joined the *Motherstone* project, we gained an articulate writer who understands the roles and values of art, spirit, and science. Through our work together, we have given the landscape a voice.

I have worked hard here to make images that tell the story of both the past and of the future as I met them in the present. They tell how the mountains and plateaus came to be and how they were continually changed until they took on their present form. The images also tell how we all can live in harmony with the natural wonders of the world and deep within the knowledge and wisdom of a common cultural wealth, and with each other. This is what my fellow travellers and wranglers and I have made together. It is more than we could have ever made apart. I have been honoured and blessed to have worked with such companions. *Motherstone* is a celebration of all that we have inherited and all that is most worth passing on.

You are part of our team now. Now we share the task with you.

— Chris Harris

As sun continues to shine over the Pacific Ocean to the west, night settles over the Itcha volcano.